AF598038

Methods in Molecular Biology

Series Editor
John M. Walker
School of Life and Medical Sciences
University of Hertfordshire
Hatfield, Hertfordshire, AL10 9AB, UK

For further volumes:
http://www.springernature.com/series/7651

Reporter Gene Assays

Methods and Protocols

Edited by

Robert Damoiseaux

Department of Molecular and Medical Pharmacology David Geffen School of Medicine, University of California Los Angeles, Los Angeles, CA, USA; California NanoSystems Institute, University of California Los Angeles, Los Angeles, CA, USA; Johnsson Comprehensive Cancer Center David Geffen School of Medicine, University of California Los Angeles, Los Angeles, CA, USA

Samuel Hasson

Pfizer, Inc., Cambridge, MA, USA

Editors
Robert Damoiseaux
Department of Molecular and Medical Pharmacology
David Geffen School of Medicine
University of California Los Angeles
Los Angeles, CA, USA

California NanoSystems Institute
University of California Los Angeles
Los Angeles, CA, USA

Johnsson Comprehensive Cancer Center
David Geffen School of Medicine
University of California Los Angeles
Los Angeles, CA, USA

Samuel Hasson
Pfizer, Inc.
Cambridge, MA, USA

ISSN 1064-3745 ISSN 1940-6029 (electronic)
Methods in Molecular Biology
ISBN 978-1-4939-7722-2 ISBN 978-1-4939-7724-6 (eBook)
https://doi.org/10.1007/978-1-4939-7724-6

Library of Congress Control Number: 2018934432

Printed on acid-free paper

This Humana Press imprint is published by the registered company Springer Science+Business Media, LLC, part of Springer Nature.
The registered company address is: 233 Spring Street, New York, NY 10013, U.S.A.

Foreword

The isolation, recruitment, and optimization of fluorescent proteins and luminescent enzymes found in nature to serve as reporters for bioassays have provided a vast toolbox for designing robust and sensitive assays. Studies on the jellyfish *Aequorea Victoria* revealed that this organism uses bioluminescent resonance energy transfer (BRET) between the calcium-activated protein aequorin and green fluorescent protein (GFP) to produce brilliant bioluminescence. Similarly, the sea pansy *Renilla reniformis* expresses *Renilla* luciferase (RLuc) and GFP to support bioluminescence. Aequorin has been adapted to construct sensitive assays for G protein-coupled receptors (GPCRs) through measuring calcium signaling events. Both RLuc and GFP have been isolated and optimized for use in cell-based assays to report on either transcriptional events in a so-called reporter gene assay (RGA) or post-translational events such as measuring protein-protein interactions in cells using BRET with optimized RLuc and GFP tags to the proteins of interest. Isolation and characterization of GFP and the widely used luciferase reporter enzyme—firefly luciferase (FLuc; derived from *Photinus pyralis*)—began in the 1960s–1970s, with the production of recombinant proteins and expression in mammalian cells occurring in the 1980s and 1990s. An early study on FLuc in 1947 revealed the ATP dependence of this enzyme and suggested the employment of FLuc to report on the ATP levels of samples. Further characterization of the luciferin (from the Latin lucifer, "light-bringer") substrates that are used by marine and beetle luciferases led to the formulation of optimal detection reagents. Adapting these assay-ready proteins to construct biological assays has greatly enabled molecular and chemical biology research.

Today, production of fluorescence and luminescence assay signals can be achieved by a variety of optimized enzymes and proteins. Since the first mutation was introduced to improve the fluorescence properties of GFP, described by Roger Tsein in 1995, protein engineering efforts have led to the description of nearly one hundred fluorescent proteins covering the visible spectrum as well as the infrared to enable both in vitro and in vivo assays. The 2008 Nobel Prize in Chemistry was awarded for the discovery and development of GFP as a tool to study biology, which was led by Osamu Shimomura, Martin Chalfie, and Roger Tsein. Certain enzymes have been adapted for high-throughput screening (HTS) assays through synthesis of fluorescence and luminescent substrates. For example, once a FRET-based substrate for the bacterial enzyme β-lactamase was made available, this enzyme became one of the most widely used reporter enzymes for developing and implementing cell-based assays in HTS. Additional enzymes such as secreted alkaline phosphatase (SEAP) and β-galactosidase have been adapted for HTS applications through synthesis of chemiluminescent substrates. Scientists at Promega pioneered the application of both FLuc and RLuc to cell-based assays as well as developing green and red emitting luciferases derived from the luminous click beetle *Pyrophorus plagiophthalamus.* All these luciferases are widely employed in cell-based assays aimed at HTS of large compound libraries. Luciferases were initially applied in molecular biology applications and later applied in HTS assays to drive drug discovery through the screening of large chemical libraries because the RGA format provides a readily implemented assay protocol suitable for automated screening systems. The recognition and characterization of assay interferences that occur when screening these reporters against compound libraries has fueled the development of more optimized reporters and protocols. For example, the recent introduction of small bright luciferases such as NanoLuc (derived from the deep sea shrimp *Oplophorus gracilirostris*, Promega) and TurboLuc (derived from the *Metridia* family of marine copepods, Thermo Fisher) has led to improved interpretation of RGA results using optimized protocols such as the coincident RGA format where two orthogonal luciferases (e.g., NanoLuc and FLuc) are co-expressed to report on the biology of interest. These sensitive reporters can be coupled with modern gene editing methods to support robust cell-based assays and application to HTS.

This book in the Methods in Molecular Biology series provides original chapters that demonstrate modern employment of reporter enzymes to understand biology. Included are chapters which consider ways to introduce reporter constructs into cells such as transient transfection techniques applied to functional genomics studies as well as applications of viral vectors, construction of minigenes to monitor splicing events, use of multicistronic constructs, and introduction of reporters at the endogenous locus with gene editing approaches. Reporters such as FLuc can be assayed in either live cells or using lytic protocols, and this book contains chapters covering live cell reporter gene formats to measure the mammalian cellular clock as well as the practical formulation and use of luciferase detection reagents. Scaling these technologies to HTS on automated systems comes with additional considerations. For example, both genome-wide genetic screens and high content imaging approaches need the appropriate infrastructure to handle the large amount of data that is generated in these experiments. Chapters covering project and data management are also presented within this book. Taken together, these chapters describe the wide-ranging applications and impact of reporter proteins in understanding basic biology and chemical biology.

Novartis Institutes for Biomedical Research ***Douglas Auld***
Chemical Biology and Therapeutics
Cambridge, MA, USA

Preface

Genes are the blueprints of life, using a shared genetic code to encode the diverse proteomes that facilitate cellular function. In order to understand specific gene expression patterns underlying these cellular functions, our ability to visualize and measure gene expression is exceedingly important. Reporter genes offer a fundamental insight into living organisms, and this quantitative technology has transformed our understanding of biology in health and disease. While fundamentally simple, reporter genes act as a genetically encoded beacon in the cell, facilitating a range of investigations from the measurement of NFkB activation in immune cells to high-throughput drug screens to identify modulators of pathogenic protein expression. What fascinated us about reporter genes was the creativity that this concept has inspired in countless scientists. Whether it is the form of the reporter itself as a constantly evolving toolbox including fluorescent, luminescent, and colorimetric proteins, or the background in which one applies the reporter (e.g., cultured cells or live animals), creativity pervades the history of this methodology. When we were invited to edit this first edition of *Reporter Gene Assays*, we found the task quite daunting. With so much fundamental and applied science utilizing the methodology, where would one start? How would we do justice to the recent developments such as gene editing and genome-wide screening approaches while enabling the newcomer with practical insight to adapt these technologies? Thankfully, we found a great amount of guidance from our own mentors to shape a book that would offer a wide degree of practical utility to the field. Overall, we wanted readers to learn through examples that represent the past, present, and future of innovation in reporter gene methodologies. Moreover, we structured the book to facilitate large-scale approaches by building a conceptual framework of techniques and skills not normally described in the literature. Essential to reaching this goal, we are exceedingly grateful to all of the chapter authors for their contributions. Both of us have been fortunate to have had experts in a range of reporter gene applications contribute to make this book a lasting compendium. Finally, we also wanted to thank the team at Springer led by John Walker, the MiMB series editor, for their help and support.

Cambridge, MA, USA — *Samuel Hasson*
Los Angeles, CA, USA — *Robert Damoiseaux*

Contents

Contributors

RAYMOND L. BARNHILL • *Department of Pathology, Institut Curie, University of Paris René Descartes, Paris, France*

LAURENT A. BENTOLILA • *California NanoSystems Institute, University of California Los Angeles, Los Angeles, CA, USA; Department of Chemistry and Biochemistry, University of California Los Angeles, Los Angeles, CA, USA*

NATHAN Y. BENTOLILA • *Genesis Innovation Lab, YULA School, Los Angeles, CA, USA*

VICTOR W. VAN BEUSECHEM • *RNA Interference Functional Oncogenomics Laboratory, Department of Medical Oncology, Cancer Center Amsterdam, VU University Medical Center, Amsterdam, The Netherlands*

IAN K. BLABY • *Department of Biology, Brookhaven National Laboratory, Upton, NY, USA*

CRYSTEN E. BLABY-HAAS • *Department of Biology, Brookhaven National Laboratory, Upton, NY, USA*

BARRY A. BUNIN • *Collaborative Drug Discovery, Inc., Burlingame, CA, USA*

ALEX M. CLARK • *Collaborative Drug Discovery, Inc., Burlingame, CA, USA; Molecular Materials Informatics, Inc., Montreal, QC, Canada*

ERICA COOK • *Lead Discovery and Optimization, Bristol Myers Squibb, Pennington, NJ, USA*

ROBERT DAMOISEAUX • *Department of Molecular and Medical Pharmacology, David Geffen School of Medicine, University of California Los Angeles, Los Angeles, CA, USA; California NanoSystems Institute, University of California Los Angeles, Los Angeles, CA, USA; Johnsson Comprehensive Cancer Center, David Geffen School of Medicine, University of California Los Angeles, Los Angeles, CA, USA*

KRISTINA DOLE • *Collaborative Drug Discovery, Inc., Burlingame, CA, USA*

PATRICIA DRANCHAK • *National Center for Advancing Translational Sciences, National Institutes of Health, Rockville, MD, USA*

SEAN EKINS • *Collaborations Pharmaceuticals, Inc., Raleigh, NC, USA*

KELLAN GREGORY • *Collaborative Drug Discovery, Inc., Burlingame, CA, USA*

JEFFREY HERMES • *Screening and Translational Enzymology, Roche, Basel, Canton of Basel-Stadt, Switzerland*

JAMES INGLESE • *National Center for Advancing Translational Sciences, National Institutes of Health, Rockville, MD, USA; National Human Genome Research Institute, National Institutes of Health, Bethesda, MD, USA*

ALISSA KEEGAN • *Department of Biological Chemistry, David Geffen School of Medicine, University of California Los Angeles, Los Angeles, CA, USA; Jonsson Comprehensive Cancer Center, David Geffen School of Medicine, University of California Los Angeles, Los Angeles, CA, USA; Molecular Biology Institute, David Geffen School of Medicine, University of California Los Angeles, Los Angeles, CA, USA; Eli and Edythe Broad Center of Regenerative Medicine and Stem Cell Research, David Geffen School of Medicine, University of California Los Angeles, Los Angeles, CA, USA*

JING LI • *Screening & Protein Sciences, Merck Research Labs, Merck & Co., Inc., North Wales, PA, USA; Screening & Protein Sciences, Merck Research Labs, Merck & Co., Inc., West Point, PA, USA*

NADIA K. LITTERMAN • *Collaborative Drug Discovery, Inc., Burlingame, CA, USA*

ANDREW C. LIU • *Department of Biological Sciences, The University of Memphis, Memphis, TN, USA; Department of Physiology and Functional Genomics, University of Florida College of Medicine, Gainesville, FL, USA*

YAPING LIU • *Screening & Protein Sciences, Merck Research Labs, Merck & Co., Inc., West Point, PA, USA*

CAMILA LOPEZ-ANIDO • *Waisman Center, University of Wisconsin, Madison, WI, USA*

CLAIRE LUGASSY • *Department of Translational Research, Institut Curie, Paris, France*

GARRY A. LUKE • *Biomedical Sciences Research Complex, School of Biology, University of St Andrews, Fife, Scotland, UK*

RYAN MACARTHUR • *National Center for Advancing Translational Sciences, National Institutes of Health, Rockville, MD, USA*

ANDREW M. MCNUTT • *Collaborative Drug Discovery, Inc., Burlingame, CA, USA*

SABEEHA S. MERCHANT • *Department of Chemistry and Biochemistry, University of California Los Angeles, Los Angeles, CA, USA; Institute for Genomics and Proteomics, University of California Los Angeles, Los Angeles, CA, USA*

LOREN MIRAGLIA • *Department of Genomics, The Genomics Institute of the Novartis Research Foundation, San Diego, CA, USA; Functional Genomics Screening Team, The Genomics Institute of the Novartis Research Foundation, San Diego, CA, USA*

JOHN J. MORAN • *Waisman Center, University of Wisconsin, Madison, WI, USA*

M. DUDLEY PAGE • *Department of Chemistry and Biochemistry, University of California Los Angeles, Los Angeles, CA, USA*

KATHRIN PLATH • *Department of Biological Chemistry, David Geffen School of Medicine, University of California Los Angeles, Los Angeles, CA, USA; Jonsson Comprehensive Cancer Center, David Geffen School of Medicine, University of California Los Angeles, Los Angeles, CA, USA; Molecular Biology Institute, David Geffen School of Medicine, University of California Los Angeles, Los Angeles, CA, USA; Eli and Edythe Broad Center of Regenerative Medicine and Stem Cell Research, David Geffen School of Medicine, University of California Los Angeles, Los Angeles, CA, USA*

CHIDAMBARAM RAMANATHAN • *Department of Biological Sciences, The University of Memphis, Memphis, TN, USA*

MARTIN D. RYAN • *Biomedical Sciences Research Complex, School of Biology, University of St Andrews, Fife, Scotland, UK*

ELLEN SIEBRING-VAN OLST • *Department of Pulmonary Diseases, Cancer Center Amsterdam, VU University Medical Center, Amsterdam, The Netherlands*

ANNA COULON SPEKTOR • *Collaborative Drug Discovery, Inc., Burlingame, CA, USA*

CHERYL STORK • *Division of Biomedical Sciences, University of California, Riverside, Riverside, CA, USA*

JOHN SVAREN • *Waisman Center, University of Wisconsin, Madison, WI, USA; Department of Comparative Biosciences, University of Wisconsin, Madison, WI, USA*

MATTHEW TUDOR • *Screening & Protein Sciences, Merck Research Labs, Merck & Co., Inc., North Wales, PA, USA; Screening & Protein Sciences, Merck Research Labs, Merck & Co., Inc., West Point, PA, USA*

CHARLIE WEATHERALL • *Collaborative Drug Discovery, Inc., Burlingame, CA, USA*

GENEVIEVE WELCH • *Department of Genomics, The Genomics Institute of the Novartis Research Foundation, San Diego, CA, USA; Functional Genomics Screening Team, The Genomics Institute of the Novartis Research Foundation, San Diego, CA, USA*

SIKA ZHENG • *Division of Biomedical Sciences, University of California Riverside, Riverside, CA, USA*

Chapter 1

Genome-Edited Cell Lines for High-Throughput Screening

Patricia Dranchak, John J. Moran, Ryan MacArthur, Camila Lopez-Anido, James Inglese, and John Svaren

Abstract

Measurement of gene expression for high-throughput screening is an increasingly used technique that has been developed for not only gene dosage disorders resulting from disease-associated copy number variations, but also for induction/repression of genes modulating the severity of a disease phenotype. Traditional methods have employed transient or stable transfection of reporter constructs in which a single reporter is driven by selected regulatory elements from the candidate gene. However, individual regulatory elements are inherently unable to capture the integrated regulation of multiple enhancers at the endogenous locus, and random reporter insertion can result in neighborhood effects that impact the physiological responsiveness of the reporter. Therefore, we outline a general method of employing genome editing to insert reporters into the 3′ UTR of a candidate gene, which has been used successfully in our studies of the *Pmp22* gene associated with Charcot–Marie–Tooth disease. The method employs genome editing to insert two nonhomologous reporters that maximize the efficiency of identification of biologically active molecules through concordant responses in small molecule screening. We include a number of aspects of the design and construction of these reporter assays that will be applicable to creation of similar assays in a variety of cell types.

Key words Genomics, Gene expression, HTS, Rare disease, Reporter gene, Transcription

1 Introduction

Since the development of luciferase assays that enabled rapid readout of gene expression with a high dynamic range [1], they have been used in a wide variety of screening experiments. When employed in small molecule screening assays, luciferase constructs were often generated with fusions to candidate gene promoters and/or regulatory elements [2, 3] While this strategy is informative, it is common for genes to be regulated by multiple enhancers [4], which are often generally dispersed at locations quite far away from the promoter, either upstream, downstream or within introns [5–9]. Given that selection of candidate regulatory elements is often based on incomplete information, it would be advantageous to embed reporter genes within the candidate locus. Figure 1

Robert Damoiseaux and Samuel Hasson (eds.), *Reporter Gene Assays: Methods and Protocols*, Methods in Molecular Biology, vol. 1755, https://doi.org/10.1007/978-1-4939-7724-6_1,

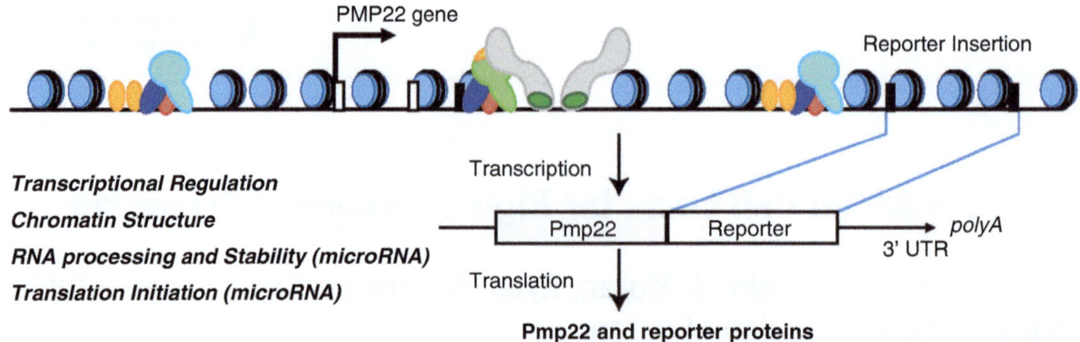

Fig. 1 The diagram shows the chromatin structure of the *Pmp22* gene including nucleosomes (circles) and transcription factors that determine *Pmp22* expression level. Since this structure cannot be recapitulated in episomal or inserted constructs, the endogenous locus provides the best insertion point for reporters. Insertion of a reporter at the 3′ end of the ORF allows reporter expression that is subject to multiple levels of regulation, including transcription and chromatin effects. In addition, retention of the 3′ UTR and the cotranslation of the reporter from the single native start codon make the single transcript subject to microRNA and other regulators of RNA processing, stability, and translation

depicts the nucleosomal structure and transcription factor binding to a typical gene. Insertion of a reporter within the native locus allows for cotranscription and cotranslation of the reporter along with the endogenous gene. This type of assay will not only capture transcriptional and epigenomic mechanisms of gene regulation, but can also be used to identify compounds that affect gene regulation at the level of RNA stability and translation initiation, both of which are subject to multiple control mechanisms including microRNA-mediated regulation.

Charcot–Marie–Tooth disease (CMT) is a class of inherited neuropathies affecting the peripheral nervous system. Over 50% of CMT cases fall within subtype 1A, which is caused by a 1.4 Mb duplication encompassing the *PMP22* (peripheral myelin protein 22) gene. In the interest of identifying modulators of *PMP22* expression, we have developed methods for reporter based screening at this locus. In these studies, we recently found that use of randomly integrated reporters in a high-throughput screen [10] utilizing a single enhancer [6] fused to a luciferase reporter were incapable of detecting some pharmacologically relevant compounds that were subsequently identified through a genome editing-enabled screen using reporters targeted to the *Pmp22* gene locus [11]. While reporters have often been inserted in endogenous loci by traditional homologous recombination techniques in mouse embryonic stem cells, the advent of genome-editing techniques [12, 13] enabled facile insertion of reporters at desired loci in a wide range of cell types [11, 14, 15]. Recent studies have used similar methods for development of other screening lines [14, 15].

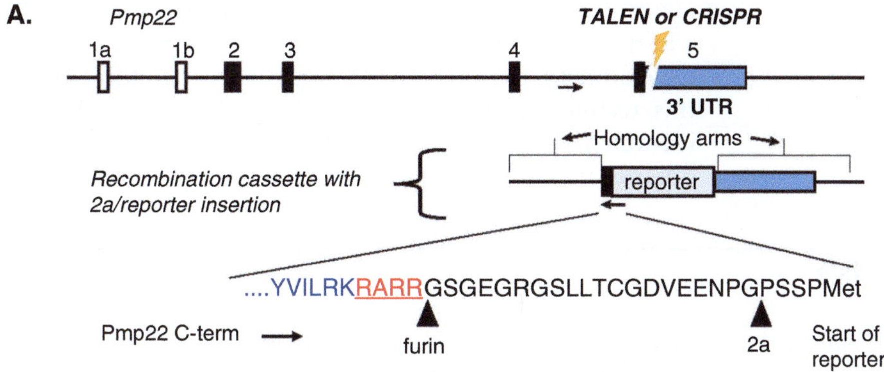

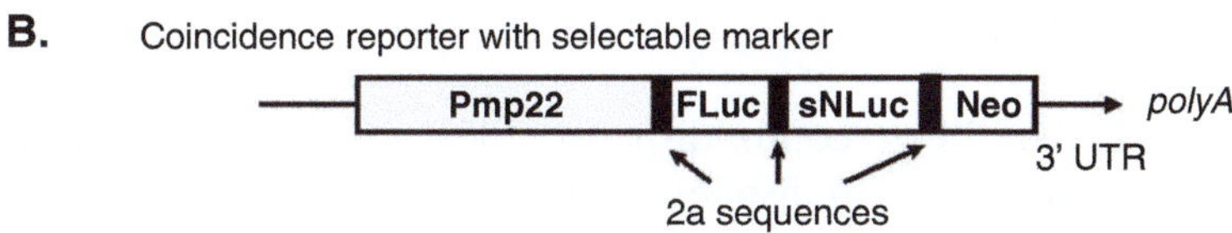

Fig. 2 (**a**) Reporter insertion is performed by genome editing in which TALEN or CRISPR constructs direct a double strand break at the 3′ end of the ORF, prompting recombination with the cotransfected cassette containing homology arms and the desired reporter(s). The 2a ribosome stuttering sequence is included at the junction of the native PMP22 protein and the reporter, and the site of 2a cleavage is indicated. In some cases, a furin cleavage site can be inserted to prevent addition of the 2a peptide at the end of PMP22. (**b**) One potential reporter combination is shown in which two coincidence reporters (firefly and secreted nanoluciferase) are inserted along with a Neomycin resistance gene, which is used to select for the desired recombinants. All four reading frames are separated by 2a ribosome stuttering sequences

Here, we outline a method to develop genome-edited reporters that are placed before the 3′ UTR of the desired, candidate gene [11, 14]. This design includes several key features. First, the "ribosome stuttering" 2a sequence [16, 17] is used between the candidate gene ORF and the reporter gene(s) to ensure that two separate proteins are coded by the same transcript (Fig. 2). Second, we employ the previously described coincidence reporter design [18], in which two unrelated luciferase reporters are placed in series behind the candidate gene ORF with an intervening 2a sequence. Incorporation of the coincidence reporter design eliminates biased selection of luciferase enzymatic modulators. Since insertion of two reporters reduces the efficiency of homologous recombination, we discuss options to enhance the efficiency of screening larger numbers of clones to obtain the desired reporter lines.

Our application of this technology has been based on a copy number variant (CNV) disorder, and the dramatic increase in human genome sequencing has uncovered a large number of CNV-associated disorders where an under- or over-expression of a gene causes a genetic disorder [19, 20]. This approach is therefore widely applicable. Moreover, the development of projects to identify genetic modifiers of disease progression [21] provides additional targets that could be the substrate of reporter assay development for therapeutic discovery. Therefore, application of this technology where specific druggable targets have not been identified can provide a means to explore the diverse chemical libraries that are available for HTS.

2 Materials

2.1 qHTS Materials

1. 1536-well white solid bottom microtiter plates (Greiner BioOne) for luminescence assays.
2. Weighted metal lids with gas exchange holes (Wako Automation).

2.2 Liquid Dispensers

1. Multidrop combi dispenser (Thermo Fisher Scientific), a peristaltic pump dispenser for cell dispensing.
2. Pintool dispenser (Kalypsis, Wako Automation), a 1536-well pin-based compound transfer tool compatible with DMSO solvent, or equivalent dispenser capable of 20–25 nL compound transfer.
3. BioRaptr flying reagent dispenser (FRD) (Beckman Coulter), a piezoelectric liquid dispenser, or multidrop combi peristaltic pump dispenser for addition of luminescence reagents.

2.3 Luminescence Reader

1. ViewLux (PerkinElmer) CCD camera-based plate reader.

2.4 Assay Medium

1. DMEM high glucose non-phenol red media with L-glutamine, 10% FBS, 1% penicillin–streptomycin in 0.9% NaCl antibiotic reagent.

2.5 Control Compounds

1. PTC124 dissolved in 100% DMSO to a stock concentration of 10 mM, Stock solution snap-frozen and stored at −20 °C.
2. Cilnidipine dissolved in 100% DMSO to a stock concentration of 20 mM. Stock solution snap-frozen and stored at −20 °C.
3. Digitonin dissolved in 100% DMSO to a stock concentration of 20 mM. Stock solution snap-frozen and stored at −20 °C.

2.6 Luminescence Reagents

1. Nano-Glo Dual-Luciferase Reporter Assay System (Promega): One-Glo Ex Luciferase reagent for FLuc read, NanoDLR Stop & Glo reagent for NLuc read. Reagents prepared according to manufacturer's protocol and protected from light. One-Glo reagent aliquoted into 10 mL tubes and stored at −20 °C. NanoDLR Stop & Glo reagent prepared fresh before each experiment. Both reagents should be filtered through 0.22 μm filter before being dispensed through a piezoelectric dispenser.

2. CellTiter-Glo (Promega) cytotoxicity reagent. Reagent prepared according to manufacturer's protocol and protected from light. CellTiter-Glo should be filtered through 0.22 μm filter before being dispensed through a piezoelectric dispenser.

3 Methods

3.1 Construction of Transgene

1. The transgene consists of two arms of homology flanking the insertion site at the 3′ end of the candidate gene's ORF (Fig. 2). One arm of homology will be derived in part from the last coding exon of the transcript and the downstream homology arm will be in the 3′ UTR so that it retains potential microRNA targeting sites. Amplify the two homology arms of approximately 800 nucleotides each. Ideally, amplification should use genomic DNA from the cell type of interest, since cell lines may contain SNPs that differ from the reference genome. It is recommended to introduce a few silent mutations within the guide RNA recognition sequence in the recombination donor construct to prevent recleavage by the CRISPR-Cas9 system after recombination.
2. The two homology arms are then assembled with the reporters (Fig. 2) using standard or Gibson cloning. Multiple reporter configurations are possible, and Table 1 includes a list of reporter cassettes that we have developed. Further description of these configurations is included below in **Notes 2–5**. In all cases, the reporter cassette contains the 2a ribosome stuttering sequence (*see* **Note 1**) cloned in frame between the native ORF and the first firefly luciferase reporter (Fig. 2). If multiple reporters are used, another 2a sequence is cloned in frame between the firefly luciferase ORF and the second reporter (e.g., nanoluciferase, *see* **Note 2**). It is important that the constructs be cloned such that the 2a sequences replace the stop codon of the candidate gene, as well as the first reporter stop codon. Cloning software can be used to verify that there is an uninterrupted ORF from the start codon of the candidate gene to the stop codon of the second reporter. Since the genome editing insertion of a transgene into the locus can be inefficient depending on the target cell type, a third gene (with another preceding 2a sequence) is cloned in series (*see* **Notes 3** and **4**). This can be a fluorescent or selectable marker (e.g., GFP or Neomycin resistance).
3. Upon cloning of the transgene, sequence the construct thoroughly to avoid cloning of PCR-induced mutations particularly in homology arms. Simultaneous sequencing of PCR fragments from the target cell line will determine if changes from the reference genome are SNPs or PCR-induced errors.

Table 1
Reporter cassette configurations for genome editing insertion at 3′ UTR

Configuration of reporter insert	Properties
2a-Gluc *	Secreted Gaussia luciferase reporter
2a-secNLuc-2a-GFP *	Secreted nanoluciferase with GFP for cell sorting
2a-NLuc-2a-GFP	Nanoluciferase with GFP
2a-NLuc-2a-puro	Nanoluciferase with puro for selection
2a-FLuc-2a-NLuc	Coincidence reporter with firefly and nanoluciferase
2a-FLuc-2a-secNLuc	Coincidence reporter with firefly and secreted nanoluciferase
2a-GFP-2a-FLuc-2a-NLuc	Coincidence reporter with GFP for cell sorting
2a-GFP-2a-FLuc-2a-secNLuc	Coincidence reporter with GFP for cell sorting
2a-FLuc-2a-Nluc-2a-Neo	Coincidence reporter with firefly luciferase, nanoluciferase and Neomycin resistance
2a-FLuc-2a-secNluc-2a-Neo	Coincidence reporter with firefly luciferase, secreted nanoluciferase, and neomycin resistance

The table lists specific combinations of reporters that can be inserted at 3′ end of target ORF. Those designated with an asterisk have been published [11], and all cassettes have been assembled in specific clones. In each case, the stop codon at the 3′ end of the ORF is designed to be replaced by an in-frame insertion of the 2a ribosome stuttering sequence, followed by one or more reporters, as in Fig. 2. Second and third genes in series are always preceded by the 2a sequence. Cassettes include both secreted (Gluc, secNluc) and nonsecreted reporters, and selection is enabled by GFP by FACS or by selectable markers (puromycin or neomycin resistance). Last two reporter cassette configurations have been used successfully (data not shown)

3.2 Transfection of Cell Line

1. Since there are a variety of transfection methods as well as genome editing techniques, they will not be specified here, but is important to use a high efficiency transfection method (unless a good selection marker is employed). In our case, the reporter transgene was cotransfected with DNA plasmids for either TALENs, or CRISPR-Cas9 [11]. RNA transfection of either genome editing reagent is also possible. It is generally recommended to target a double strand break as close as possible to the native stop codon, although it should be noted that the availability of high quality target sites for CRISPR-Cas9 may affect design of the desired recombination.
2. At 1–4 days after transfection, cells can be prepared for FACS sorting of fluorescent marker (*see* **Note 3**) or antibiotic selection can be applied (*see* **Note 4**). For antibiotics, ensure that the appropriate level of antibiotic has been established for the cell line. For antibiotics that take more time for cell death (e.g., G418), ensure that cells do not become prematurely confluent

during selection. For FACS, fluorescent cells can be sorted as single cells into a 96-well plate. Some cell lines do not grow well under such conditions, so conditioned medium can be used to enhance survival of single cells.

3.3 Identification and Evaluation of Clones

1. Evaluation of candidate clones should include use of known modulators of the candidate gene (either small molecule or siRNA) to show that the reporter assays are similarly regulated as the candidate gene. For example, we employed siRNA for a known transcriptional regulator of *Pmp22*, Sox10 [6], to determine that the reporter activity responded in a similar manner to the endogenous gene [11]. Known pharmacological modulators can be similarly employed. If a secreted reporter is used, it is important to wash the medium at 24–48 h after siRNA transfection, and then allow reporter accumulation for 30–60 min prior to reporter measurement (*see* **Note 5**).
2. Isolate genomic DNA from the clones and perform PCR with at least one primer outside of the homology arm and another in the 2a sequence (or reporter) to test that the reporters are indeed inserted at the endogenous gene locus. Due to the inherent inefficiency in inserting reporter cassettes, it is anticipated that most positive lines will have a single modified allele, but this can be tested by PCR for retention of the wild-type allele in the reporter clones.
3. Another caveat is that the cloning process may uncover clones that differ substantially from the parental cell line due to selection pressures in the cloning process. Therefore, some evaluation of morphology and limited gene profiling of cell type-specific genes by qRT-PCR should be used to assess how well the reporter line resembles the parental cell line. For diploid cell lines, it is prudent to also establish that karyotype is unaltered.
4. The 2a stuttering sequence is generally quite efficient, but a Western blot for the protein of the candidate gene can be used to validate that the original protein is not fused to one or more of the reporters.

3.4 Miniaturization and qHTS Compatibility Validation of Clones

1. Optimization and validation of individual clones for evaluation in 1536-well microtiter plate format is necessary prior to any small molecule or biologic screening effort. Optimal cell density should be determined by plating various numbers of cells per well across a 1536-well plate. To determine the minimal number of cells required to obtain an acceptable and cost-effective assay output when scaling up for HTS, 500–1500 cells/well is a recommended cell density range to test. Four different cell densities are plated in quadrants at 4 μL per well across white solid-bottom 1536-well plates in fresh assay medium with a multidrop combi dispenser (*see* **Note 6**).

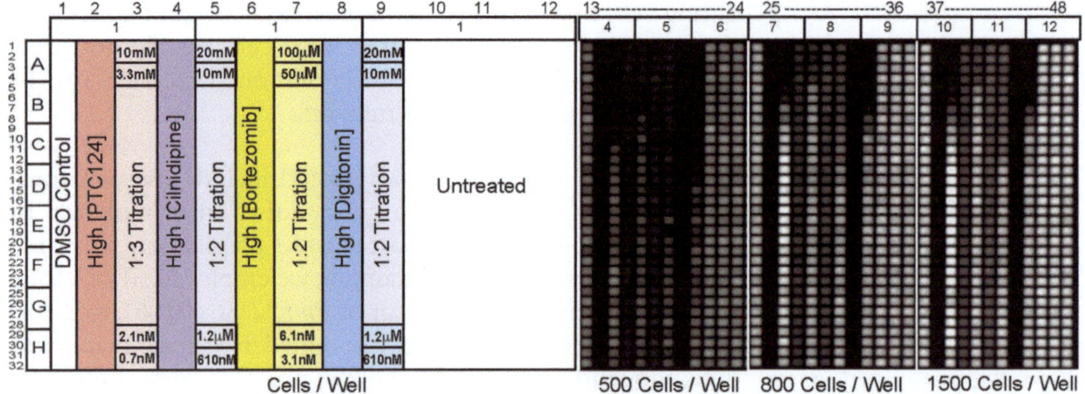

Fig. 3 1536-well plate layout for cell density and control compound optimization. Cells plated at four different cell densities across quadrants of a 1536-well white solid-bottom, plate. Control compounds at high concentration and 16-pt titrations prepared and added to cells with pintool transfer of 23 nL per well. DMSO vehicle control, PTC124 used as FLuc reporter control, Cilnidipine used as NLuc reporter control, Bortezomib used as biological pathway control, and Digitonin used as cytotoxicity control. (Right side) FLuc luminescence plate image of cell density and control compound optimization from ViewLux CCD-based plate reader. Increased luminescence is observed as cell density increases. At low cell densities increased noise, well to well variability, is observed. In loss of signal assays, decreased luminescence (darker wells) is observed in columns treated with respective controls. Titration responses are observed with various potencies between control compounds but consistent titrations across cell densities. Titration of PTC124 in columns 15, 27, and 39 demonstrate the expected inhibition of FLuc enzyme at high concentrations of the compound followed by enzyme stabilization at moderate concentrations indicating appropriate integration of the reporter gene in the cells. Viewlux settings: 30 s exposure, high gain, slow speed, 2× binning. Image is autoscaled based on highest and lowest signal on the plate

2. In addition to determining optimal cell density for basal reporter signal, each cell density for individual clones should be tested across pharmacological controls, if available, as well as reporter control compounds (*see* ref. 14 for examples). Each compound should be tested in at least 32 replicates (one column of 1536-well plate) at the concentration that achieves a maximum response (e.g., 3–10-fold above the EC/IC_{50}) to allow for the determination of the response range and assay statistics [22, 23], and tested in a 16-pt, 1:2 or 1:3 titration in duplicate per cell density. DMSO vehicle control should also be included in at least 32 replicates (Fig. 3). Twenty-three nanoliters of respective control compound is added to each assay well by pintool transfer resulting in a final 174-fold dilution. PTC124 and Cilnidipine are useful FLuc and NLuc control compounds, respectively, which directly bind and modulate the reporter enzyme half-life in the cell [6]. PTC124 is tested at a high concentration of 57 μM and titrated in a 16-pt, 1:3 dilution resulting in a final concentration range of 4 pM–57.5 μM, while Cilnidipine is used at a high concentration of 115 μM and titrated in a 16-pt, 1:2 dilution resulting in

a final concentration range of 3.5 nM–115 μM. Some cytotoxicity may be observed in certain cell lines at the high concentration of cilnidipine. Digitonin can be used as a cytotoxicity control across most cell lines at a high concentration of 57.5 μM and titrated in a 16-pt 1:2 dilution resulting in a final concentration range of 1.8 nM–57.5 μM. Compounds should be added to cells within an hour of plating if a secreted reporter is used to minimize the basal accumulation of reporter present in the media and to maximize signal to background for a given assay. If nonsecreted reporters are used, cells can be incubated 16–24 h at 37 °C, 5% CO_2, and 95% RH prior to compound addition to allow the cells to recover from the stress of plating. Cells should be incubated at 95% RH with weighted metal lids with gas exchange holes to minimize evaporation and edge effects that can occur in 1536-well plates (*see* **Note** 7).

3. Depending on the biology of interest, cells should be incubated with compound 8–48 h at 37 °C, 5% CO_2, and 95% RH. A time course for optimal response can be performed by plating and treating replicate plates above that are assayed at various time points. FLuc and NLuc coincidence luminescence are measured with the Nano-Glo Dual-Luciferase reporter assay system according to the manufacturer's protocol. All reagents should be filtered through a 0.22 μm membrane to remove particulates prior to dispensing through a piezoelectric or peristaltic pump to prevent clogging of the tips and ensure even distribution of the reagent. In short, one volume of One-Glo EX luciferase reagent, which can be reconstituted and stored at −30 °C protected from light for long term storage, is added to each well with a BioRAPTR FRD or equivalent dispenser, incubated for 10 min at room temperature, and FLuc luminescence is measured on a ViewLux plate reader. One volume NanoDLR Stop & Glo reagent prepared fresh is added as above, incubated 10 min at room temperature, and NLuc luminescence is measured on a ViewLux plate reader (Table 2). Detector settings on the plate reader should be optimized for individual reporters in each assay so that no well on any given plate is over exposed, while reasonable basal luminescence signal is achieved at the lowest cell density. It is suggested to change exposure and gain rather than binning, if possible, to maintain consistent resolution in CCD camera-based systems.
4. A cytotoxicity counter screen should also be optimized for each clone prior to any significant screening effort to determine the contribution of compound cytotoxicity to the output signal. The initial pharmacological characterization of compounds identified in loss-of-signal assays, as well as bell-shaped dose response curves identified in gain-of-signal assays require an understanding of the acute cytotoxic properties of active

Table 2
Protocol table for miniaturization of genome-edited coincidence reporter cell lines and control compound optimization to 1536-well qHTS format

Protocol Optimziation Table: Generic sequential read reporter cell line			
Step	**Parameter**	**Value**	**Description**
1	Cells	4 μL	Seed 500, 800, 1000 or 1500 cells/well in one Greiner white/solid bottom high base, TC plate per genome-edited FLuc-P2A-secNLuc clone
2	Compounds	23 nL	Compound transfer by Pintool: Bortezomib (100 μM high concentration; 16-pt 1:2 titration) used as positive biological control; PTC124 (10 mM high concentration; 16-pt 1:3 titration) and Cilnidipine (20 mM high concentration; 16-pt 1:2 titration) used as reporter controls for FLuc and NLuc respectively; and Digitonin (20 mM high concentration; 16-pt 1:2 titration) used as cytotoxicity control
3	Incubation	24 h	Incubate at 37 °C, 5% CO_2, 95% RH for 24 h
4	Reagent	4 μL	Add One-Glo EX FLuc luminescence reagent (Promega) with BioRaptr FRD
5	Incubation	10 min	Incubate at room temperature, protected from light; Cell lysis
6	Output 1	Viewlux	Luminescence Read #1 (FLuc): Exposure = 30 s, Gain = High, Speed = Slow, Binning = 2×
7	Reagent	4 μL	Add NanoDLR Stop & Glo reagent (Promega) with BioRaptr FRD
8	Incubation	10 min	Incubate at room temperature, protected from light
9	Output 2	Viewlux	Luminescence Read #2 (NanoLuc): Exposure = 1 s, Gain = High, Speed = Slow, Binning = 2×
Step	**Notes**		
1	Cells plated in standard growth media. It is recommended to add 1% Penicillin-Streptomycin (10,000 U/mL) to media prior to plating.		
2	Compounds prepared at high concentration stocks in DMSO and plated with vehicle control in source 384-well plate, titrated as specified, and manually transferred in duplicate to a 1536-well polypropylene V-bottom compound plate compatible for use with pintool		
4 and 7	Prepare Nano-Glo DLR Assay Reagents: (**a**) Reconstitute NanoDLR One-Glo EX FLuc substrate by adding 100 mL Dual-Glo Luciferase buffer to one bottle of lyophilized substrate, mix well and protect from light. (May be kept at −30 °C for long term storage); (**b**) Prepare NanoDLR Stop and Glo Reagent (just enough needed for specific experiment): Dilute the NanoDLR Stop & Glo substrate 1:100 in NanoDLR Stop & Glo buffer = Add 60 μL of Dual-Glo Stop & Glo Substrate to 6 mL Stop & Glo Buffer (per 1536-well plate), mix well and protect from light (Prepare fresh for each experiment and store at room temperature, protected from light); (**c**) Filter both reagents through a 0.22 μm membrane to remove particulates		

compounds identified in HTS. For example, bell-shaped concentration–response curves (CRCs) seen in gain-of-signal assays may be due to cytotoxic effects of a given compound at the higher concentrations or a non-cytotoxic consequence of signal quenching or direct reporter enzyme inhibition [24]. Cells should be plated, treated (including a cytotoxicity control compound such as Digitonin), and incubated as above. Cytotoxicity can be measured by either fluorescence or luminescence assays. CellTiter-Glo is a common luminescence based assay that measures cellular ATP which is directly correlated to cell viability. Cells are plated and treated as above, then one volume CellTiter-Glo reagent is added to each assay well with a BioRAPTR FRD, incubated for 10 min at room temperature, and luminescence is measured on a ViewLux plate reader with optimized measurement settings for a given cellular assay system (Fig. 3) (*see* **Note 8**).

5. Quantified luminescence values for each variable optimized above should be analyzed for basic HTS statistics. First, average signal, standard deviation, and coefficient of variation (CV) should be calculated for the high concentration replicates of each of the control compounds, as well as DMSO treated and untreated wells at each cell density and time point.

$$\text{Signal to background (S : B)} = \frac{\text{AVE signal neutral control}}{\text{AVE signal positive control}} \quad (1)$$

$$\text{Signal to noise (S : N)} = \frac{\text{AVE positive control} + \text{AVE neutral control}}{\sqrt{(\text{STDEV positive control}^2 + \text{STDEV neutral control}^2)}} \quad (2)$$

$$Z' \text{ factor} = 1 - \frac{(3\text{actor} = 1 - \text{tive control} + 3 \cdot \text{STDEV neutral control})}{|\text{AVE positive control}) + \text{AVE neutral control}|} \quad (3)$$

Signal to background (S:B) (Eq. 1), signal to noise (S:N) (Eq. 2), and Z' factor (Eq. 3) statistics should be calculated for each control compound compared to DMSO vehicle control under each set of assay conditions. General guidelines for application of these statistics in the determination of optimal assay conditions include CV < 10 in DMSO vehicle control and untreated wells, and a Z' factor > 0.5 which takes into account S:B and assay signal variation (S:N) between DMSO control wells and the high concentration of a given control compound (Table 2) [22]. Additionally, concentration response curves should be plotted and AC_{50} values determined for each control compound titration. Software such as Graphpad Prism is recommended for curve fitting and AC_{50} determination for the control compound titrations for each of the assay variables tested (Fig. 4).

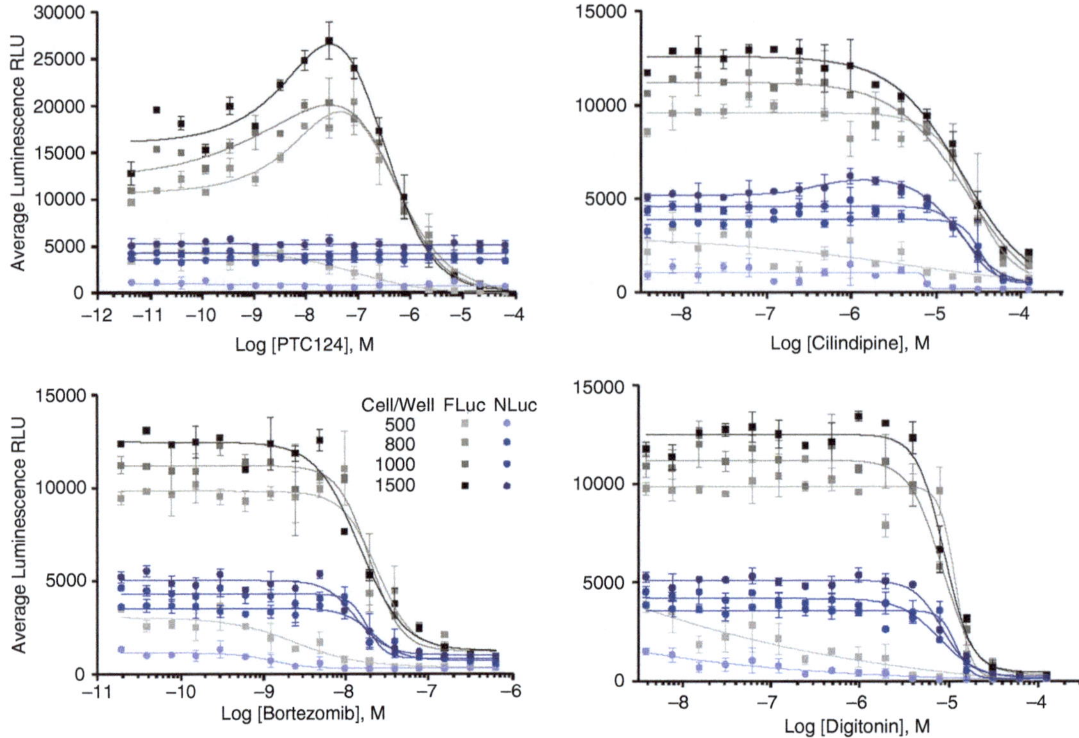

Fig. 4 Concentration response curves (CRCs) for the four control compounds tested across different cell densities for a given genome-edited FLuc-P2A-secNLuc cell line. FLuc CRCs are shown with gray symbols and NLuc CRCs are shown with blue symbols. Data was plotted and curves fit in Prism software (GraphPad). FLuc specific stabilization with PTC124 was observed as bell-shaped curves at the three highest cell densities, while NLuc stabilization was only observed with Cilnidipine at the highest cell density. Both Bortezomib, the biological control in this assay, and Digitonin, the cytotoxicity control, demonstrated consistent concordant IC_{50} values at the three highest cell densities. Based on these results 1500 cells/well would be selected as the best cell density with which to use in future qHTS. Error bars represent the standard deviation of two technical replicates

6. Once optimal assay conditions are determined based on control compound performance, a preliminary pilot quantitative high-throughput screen (qHTS) should be performed to validate assay performance. Commercially available libraries such as The Library of Pharmacologically Active Compounds ($LOPAC^{1280}$) from Sigma-Aldrich are suggested for this purpose (*see* **Note 9**). Library compounds are plated in 384-well plates comprising of seven interplate titrations series (serial dilutions of 1:5 in DMSO) for concentrations ranging from 10 mM to 0.64 μM. Interplate titrations are then reformatted into 1536-well plates for compound addition to assay wells. Cells are plated at the determined optimal cell density in 4 μL per well across white solid-bottom 1536-well plates in fresh

assay medium with a multidrop combi dispenser as above. Library compounds are transferred in interplate titrations to assay plates, while appropriate biological and reporter controls are added to each assay plate as described above. It is recommended to include one column of DMSO vehicle control, 8–32 replicate wells of high concentration blocks of each control compound, and 8–16-pt titrations in duplicate of each control to the first four columns of every assay plate. Cells are incubated with compounds and assayed as above for both FLuc and NLuc luminescence at the optimal time point to observe the appropriate biological response. It is also recommended to include a DMSO control plate (a control library plate consisting of DMSO in columns 5–48, with control compounds added as with standard library plates) at the beginning and end of the screen for post-screening correction of systematic errors (*see* **Note 10**).

7. Statistical analysis and server software will be needed to process qHTS data. Screening results should be corrected for systematic errors using the *B*-score method and normalized to controls [25]. Normalized qHTS data is typically described as percent inhibition, in a loss of signal assay, or percent activation, in a gain of signal assay. Potency values for active small molecules can be determined in qHTS and presented as EC_{50}, IC_{50} or AC_{50}. Assay performance should be assessed based on the Z' factor of the controls included on each plate and the minimum significance ratio (MSR) of the intra-plate control titrations, where an MSR < 2 is considered a reproducible assay [26].

4 Notes

1. It is important to note that the 2a stuttering sequence does introduce several amino acids, and in some cases, an additional furin cleavage site cloned just upstream of the 2a sequence [17] can be used to remove the extra 2a sequence (Fig. 2).
2. A number of luciferase reporters have been identified, and even individual subtypes are available with varying half-lives, as well as secreted versions. Both firefly (FLuc) and nanoluciferase (NLuc) reporters are available in shorter half-life versions that have altered kinetics of induction and/or repression. However, it should be noted that even a highly active candidate gene will express the reporter from a single copy insertion, and our experience is that the more stable luciferase versions may be required to have adequate signal for high-throughput screening in plate luminometer formats where cell numbers per well are low (e.g., 1536-well screening) [11].

3. If the candidate gene is already expressed at a certain level in the target cell line, one can clone the selectable marker in series with the reporters (with an intervening 2a sequence) so that the marker is expressed from the same regulatory elements that drive the endogenous gene and inserted reporters. We have employed fluorescent reporters to allow isolation of candidate clones by FACS sorting and selectable markers such as neomycin and puromycin for antibiotic selection [11]. In either case, use of this method is advantageous since many of the homology arms are likely to lack regulatory elements. Therefore, random insertion of the transgene rather than insertion by homologous recombination will likely not provide adequate expression levels of either fluorescent or antibiotic selection markers.
4. In some cases the candidate gene may not be expressed at very high levels, and therefore, an independent cassette can be cloned into the construct. One such example is the commonly used PGK-Neo cassette, which can be flanked by loxP sites. A selection marker can be used to identify desired clones and cre-mediated recombination can then be used to remove the marker as needed.
5. Another design consideration is selection of a secreted luciferase, such as Gaussia luciferase (GLuc) [27] or secreted nanoluciferase (secNLuc) [28]. Since secreted enzymes are quite stable, this design feature provides a significant boost in signal compared to nonsecreted versions. However, the accumulation of stable reporter in the medium is also a problem, particularly in screening for compounds that reduce gene expression. In practical terms, this can be circumvented by washing the cells at some point after compound addition, and then allowing cells to secrete enzyme 30–60 min prior to measurement in order to assess accumulated enzyme levels from addition of the compound. Since this is not practical for high-throughput screening, compounds can also be administered within 1–2 h of plating to minimize reporter accumulation in the absence of compounds. Another important advantage of secreted reporters is that candidate clones can be screened without having to replicate clones, which can be quite time-consuming, as a relatively small amount of medium can be removed for reporter measurement. Therefore, for many situations, a combination of nonsecreted reporter (e.g., firefly) and a secreted reporter will have a number of advantages in identifying candidate clones.
6. When plating cells in 1536-well format the use of a peristaltic pump dispenser is suggested as it is gentler on cells than piezoelectric dispensers and more time efficient and consistent than manual plating. It is also recommended to filter the cells

through at least a 40 μm mesh size strainer prior to dispensing for uniform cell suspensions.

7. An alternative to weighted metal lids for 1536-well plates is to seal microtiter plates with breath-easy sealing membranes (Sigma-Aldrich) to allow for CO_2 gas exchange while minimizing evaporation from individual wells that can increase well-to-well variability. When breath-easy membranes are used, plates should be centrifuged prior to removing the seals to minimize spill over between wells.

8. There are several different methods by which to interrogate compound cytotoxicity in cell based assays. A nonlytic, fluorescence output alternative to the CellTiter-Glo luminescence assay described above is CellTiter-Fluor (Promega). The CellTiter-Fluor cell viability assay measures the activity of a conserved protease, where a cell permeable profluorescent substrate is cleaved by active protease in live cells generating fluorescence in direct correlation with cell number. The assay should be performed according to manufacturer's protocol where one volume of reagent is added per well and fluorescence is measured (excitation 388–400 nm, emission 505 nm) on a plate reader (such as PerkinElmer's Envision or the Tecan Infinite M1000 Pro).

9. There are several commercially available diversity libraries, including the Prestwick Chemical Library and Tocriscreen Plus collection that are comparable in size and composition to the LOPAC library. Any of these chemical libraries can be used to validate a 1536-well qHTS assay. Regardless of the chemical library tested, compounds should be transferred to assay plates from low to high concentration to minimize the effects of carry over that can occur with pintool transfer.

10. DMSO control plates placed at the beginning and end of a library screen can be used to indicate and correct for dispense errors that may occur across a screen during cell or reagent addition, as well as indicate the amount of carryover from high concentration library plates.

References

1. Thorne N, Inglese J, Auld DS (2010) Illuminating insights into firefly luciferase and other bioluminescent reporters used in chemical biology. Chem Biol 17(6):646–657. https://doi.org/10.1016/j.chembiol.2010.05.012
2. Michelini E, Cevenini L, Mezzanotte L, Coppa A, Roda A (2010) Cell-based assays: fuelling drug discovery. Anal Bioanal Chem 398(1):227–238. https://doi.org/10.1007/s00216-010-3933-z
3. Chancellor DR, Davies KE, De Moor O, Dorgan CR, Johnson PD, Lambert AG, Lawrence D, Lecci C, Maillol C, Middleton PJ, Nugent G, Poignant SD, Potter AC, Price PD, Pye RJ, Storer R, Tinsley JM, van Well R, Vickers R, Vile J, Wilkes FJ, Wilson FX, Wren SP, Wynne GM (2011) Discovery of 2-arylbenzoxazoles as upregulators of utrophin production for the treatment of Duchenne muscular dystrophy. J Med Chem 54

(9):3241–3250. https://doi.org/10.1021/jm200135z

4. Kuznetsova T, Stunnenberg HG (2016) Dynamic chromatin organization: role in development and disease. Int J Biochem Cell Biol 76:119–122. https://doi.org/10.1016/j.biocel.2016.05.006
5. Jang SW, Svaren J (2009) Induction of myelin protein zero by early growth response 2 through upstream and intragenic elements. J Biol Chem 284(30):20111–20120
6. Jones EA, Lopez-Anido C, Srinivasan R, Krueger C, Chang LW, Nagarajan R, Svaren J (2011) Regulation of the PMP22 Gene through an Intronic Enhancer. J Neurosci 31 (11):4242–4250. https://doi.org/10.1523/JNEUROSCI.5893-10.2011. 31/11/4242 (pii)
7. Jones EA, Brewer MH, Srinivasan R, Krueger C, Sun G, Charney KN, Keles S, Antonellis A, Svaren J (2012) Distal enhancers upstream of the Charcot-Marie-Tooth type 1A disease gene PMP22. Hum Mol Genet 21:1581–1591. https://doi.org/10.1093/hmg/ddr595. ddr595 (pii)
8. Bujalka H, Koenning M, Jackson S, Perreau VM, Pope B, Hay CM, Mitew S, Hill AF, QR L, Wegner M, Srinivasan R, Svaren J, Willingham M, Barres BA, Emery B (2013) MYRF is a membrane-associated transcription factor that autoproteolytically cleaves to directly activate myelin genes. PLoS Biol 11 (8):e1001625. https://doi.org/10.1371/journal.pbio.1001625
9. Hung HA, Sun G, Keles S, Svaren J (2015) Dynamic regulation of schwann cell enhancers after peripheral nerve injury. J Biol Chem 290 (11):6937–6950. https://doi.org/10.1074/jbc.M114.622878
10. Jang SW, Lopez-Anido C, MacArthur R, Svaren J, Inglese J (2012) Identification of drug modulators targeting gene-dosage disease CMT1A. ACS Chem Biol 7(7):1205–1213. https://doi.org/10.1021/cb300048d
11. Inglese J, Dranchak P, Moran JJ, Jang SW, Srinivasan R, Santiago Y, Zhang L, Guha R, Martinez N, MacArthur R, Cost GJ, Svaren J (2014) Genome editing-enabled HTS assays expand drug target pathways for Charcot-Marie-tooth disease. ACS Chem Biol 9 (11):2594–2602. https://doi.org/10.1021/cb5005492
12. Cong L, Ran FA, Cox D, Lin S, Barretto R, Habib N, Hsu PD, Wu X, Jiang W, Marraffini LA, Zhang F (2013) Multiplex genome engineering using CRISPR/Cas systems. Science 339(6121):819–823. https://doi.org/10.1126/science.1231143
13. Mali P, Yang L, Esvelt KM, Aach J, Guell M, DiCarlo JE, Norville JE, Church GM (2013) RNA-guided human genome engineering via Cas9. Science 339(6121):823–826. https://doi.org/10.1126/science.1232033
14. Hasson SA, Fogel AI, Wang C, MacArthur R, Guha R, Heman-Ackah S, Martin S, Youle RJ, Inglese J (2015) Chemogenomic profiling of endogenous PARK2 expression using a genome-edited coincidence reporter. ACS Chem Biol 10(5):1188–1197. https://doi.org/10.1021/cb5010417
15. Lang L, Ding HF, Chen X, Sun SY, Liu G, Yan C (2015) Internal ribosome entry site-based bicistronic in situ reporter assays for discovery of transcription-targeted lead compounds. Chem Biol 22(7):957–964. https://doi.org/10.1016/j.chembiol.2015.06.009
16. de Felipe P, Luke GA, Hughes LE, Gani D, Halpin C, Ryan MD (2006) E unum pluribus: multiple proteins from a self-processing polyprotein. Trends Biotechnol 24(2):68–75. https://doi.org/10.1016/j.tibtech.2005.12.006
17. Fang J, Qian JJ, Yi S, Harding TC, GH T, VanRoey M, Jooss K (2005) Stable antibody expression at therapeutic levels using the 2A peptide. Nat Biotechnol 23(5):584–590. https://doi.org/10.1038/nbt1087. nbt1087 (pii)
18. Cheng KC, Inglese J (2012) A coincidence reporter-gene system for high-throughput screening. Nat Methods 9(10):937. https://doi.org/10.1038/nmeth.2170
19. Lupski JR (2015) Structural variation mutagenesis of the human genome: impact on disease and evolution. Environ Mol Mutagen 56 (5):419–436. https://doi.org/10.1002/em.21943
20. Carvalho CM, Lupski JR (2016) Mechanisms underlying structural variant formation in genomic disorders. Nat Rev Genet 17 (4):224–238. https://doi.org/10.1038/nrg.2015.25
21. Harper AR, Nayee S, Topol EJ (2015) Protective alleles and modifier variants in human health and disease. Nat Rev Genet 16 (12):689–701. https://doi.org/10.1038/nrg4017
22. Inglese J, Johnson RL, Simeonov A, Xia M, Zheng W, Austin CP, Auld DS (2007) High-throughput screening assays for the identification of chemical probes. Nat Chem Biol 3 (8):466–479. https://doi.org/10.1038/nchembio.2007.17
23. Inglese J, Shamu CE, Guy RK (2007) Reporting data from high-throughput screening of small-molecule libraries. Nat Chem Biol 3

(8):438–441. https://doi.org/10.1038/nchembio0807-438

24. Thorne N, Auld DS, Inglese J (2010) Apparent activity in high-throughput screening: origins of compound-dependent assay interference. Curr Opin Chem Biol 14(3):315–324. https://doi.org/10.1016/j.cbpa.2010.03.020
25. Brideau C, Gunter B, Pikounis B, Liaw A (2003) Improved statistical methods for hit selection in high-throughput screening. J Biomol Screen 8(6):634–647. https://doi.org/10.1177/1087057103258285
26. Eastwood BJ, Farmen MW, Iversen PW, Craft TJ, Smallwood JK, Garbison KE, Delapp NW, Smith GF (2006) The minimum significant ratio: a statistical parameter to characterize the reproducibility of potency estimates from concentration-response assays and estimation by replicate-experiment studies. J Biomol Screen 11(3):253–261. https://doi.org/10.1177/1087057105285611
27. Tannous BA (2009) Gaussia luciferase reporter assay for monitoring biological processes in culture and in vivo. Nat Protoc 4 (4):582–591. https://doi.org/10.1038/nprot.2009.28
28. Hall MP, Unch J, Binkowski BF, Valley MP, Butler BL, Wood MG, Otto P, Zimmerman K, Vidugiris G, Machleidt T, Robers MB, Benink HA, Eggers CT, Slater MR, Meisenheimer PL, Klaubert DH, Fan F, Encell LP, Wood KV (2012) Engineered luciferase reporter from a deep sea shrimp utilizing a novel imidazopyrazinone substrate. ACS Chem Biol 7(11):1848–1857. https://doi.org/10.1021/cb3002478

Chapter 2

High-Throughput Firefly Luciferase Reporter Assays

Ellen Siebring-van Olst and Victor W. van Beusechem

Abstract

Firefly luciferase reporter gene assays find wide application in high-throughput screens to identify molecular components of biological networks or to identify chemical compounds capable of interfering with cellular signaling. Here, we present methods to prepare affordable firefly luciferase assay reagents and procedures to use these reagents in reporter gene high-throughput screening with large batches of 96-well cell culture plates.

Key words Firefly luciferase, High-throughput screening, Reporter gene assay, Assay reagent, Stable luminescence

1 Introduction

Cell-based reporter gene assays are widely used to study the regulation of signal transduction pathways. A typical strategy is to clone the coding region of a reporter gene downstream of a regulatory promoter element that is responsive to a transcription factor, such that expression of the encoded reporter protein is controlled by the transcription factor. This way, signaling cascades culminating in stimulation of an interaction of the transcription factor with the response element can be monitored. Useful reporter proteins for this purpose are those that can be easily distinguished from endogenous cellular proteins using simple detection methods. In this respect, bioluminescence reporter genes are of particular interest, because mammalian cells do not exhibit any background bioluminescence.

One of the most popular reporter proteins for studies in mammalian cells is firefly (*Photinus pyralis*) luciferase [1] (*see* **Note 1**). This bioluminescent protein catalyzes adenosine triphosphate (ATP)-dependent oxidative decarboxylation of its substrate D-luciferin to oxyluciferin with concomitant light emission [2]. Under the appropriate conditions, the amount of emitted light is linearly proportional to the luciferase protein concentration. Firefly

Robert Damoiseaux and Samuel Hasson (eds.), *Reporter Gene Assays: Methods and Protocols*, Methods in Molecular Biology, vol. 1755, https://doi.org/10.1007/978-1-4939-7724-6_2, © Springer Science+Business Media, LLC, part of Springer Nature 2018

luciferase exhibits high specific activity with a broad dynamic range, usually up to more than seven orders of magnitude [3, 4]. Furthermore, the protein is short-lived, which makes it very suitable to monitor experimental modulation of cellular signaling. Apart from its utility as a reporter gene, firefly luciferase is also increasingly being used in assays for specific enzymatic activity and protein-protein interaction. In the first case, the coding region for luciferase is fused to a blocking peptide via a specific cleavable linker [5]. In the absence of enzymatic cleavage, the fusion protein is inactive and luminescence is low. In contrast, when the enzyme capable of cleaving the linker is activated, the blocking peptide is cleaved off, releasing functional firefly luciferase starting light emission. In the second case, the firefly luciferase gene is split into two parts that encode separate protein domains that do not have catalytic activities on their own and each domain is fused to one of a pair of proteins capable of interacting with each other [6]. Under conditions that induce these proteins to physically interact, the two luciferase domains are brought together reconstituting luciferase catalytic activity and thus initiating light emission.

The natural reaction when all components for the firefly luciferase catalytic reaction are present is the rapid emission of a flash of light, which decays within minutes [7]. Newer developments in reagent composition have resulted in more stable light emission profiles, thus making firefly luciferase also attractive for use in cell-based high-throughput screening (HTS) (*see* **Note 2**). In such screens, a firefly luciferase reporter assay is used for example to identify molecular components of biological networks or to identify chemical compounds capable of interfering with cellular signaling. There is thus a great need for reliable firefly luciferase assay reagents for use in HTS.

Reagent kits for sensitive detection of luciferase activity in a homogeneous assay format with long-lasting luminescence are available from several vendors (e.g., Promega, PerkinElmer, and Applied Biosystems). As these reagents can be cost prohibitive in the volumes needed for high-throughput assays, we have developed our own custom formulation [8]. Here, we present methods to prepare firefly luciferase assay reagent (FLAR) and procedures to use the reagent in reporter gene HTS. FLAR composition and pH was optimized to balance fluorescence signal intensity and stability. To reduce assay reagent cost, we reduced the concentration of the expensive component D-luciferin. Importantly, we omitted coenzyme A, which is sometimes included in firefly luciferase assay reagents [9, 10], from FLAR because we observed that while this component boosts flash luminescence, it strongly reduces the stability of light emission (*see* **Note 3**). Conditions to perform the assay were tested in 96-well microplate format useful for HTS. Signal intensity and stability was tested over a period ranging from immediately after reagent addition until 2 h after reagent addition.

This was considered to create a window large enough to measure batches of assay plates in HTS. In these studies, we used tumor suppressor p53 wild-type human cancer cell lines stable transfected with a reporter gene plasmid carrying the firefly luciferase gene driven by a p53-dependent promoter [8]. Figure 1 shows the different workflows to perform the assay, which are described in detail below. Firefly luciferase HTS assays could conveniently be performed under homogenous conditions with minimal manipulations. In these assays, firefly luciferase was released from cells by detergent addition or by a single freeze–thaw step (*see* **Note 4**). Higher (three- to fivefold) luminescence was reached when culture medium was removed, either by performing a medium-for-buffer exchange washing procedure or by removing all culture medium by spinning culture plates upside down in a standard laboratory centrifuge (*see* **Note 5**). Procedures with medium removal are more laborious, but may be preferred when reporter gene expression is low.

2 Materials

2.1 Preparation and Storage of Stock Solutions

Use high purity chemical reagents from a reliable manufacturer. We used reagents from Sigma; except Triton X-100, which was from Bio-Rad; and D-luciferin, which was from Caliper Life Sciences (this product is currently available as XenoLight D-luciferin from PerkinElmer). Prepare all stock solutions using ultrapure water (prepared by purifying deionized water to attain a sensitivity of 18 MΩ cm at 25 °C).

1. Tricine (200 mM, 50 mL, pH 7.8). Add 1.79 g Tricine to 40 mL H_2O in a glass bottle or beaker and stir until dissolved. Correct the pH to 7.8 using HCl or NaOH. When corrected, add H_2O to a final volume of 50 mL. Store at room temperature.
2. $MgCO_3$ (10 mM, 50 mL, pH 7.8). Add 243 mg $(MgCO_3)_4\cdots Mg(OH)_2\cdots 5H_2O$ (magnesium carbonate hydroxide pentahydrate) to 40 mL H_2O in a glass bottle or beaker and stir while measuring pH. Add 10 N HCl until pH is below 4.0. Keep the pH at 4.0 until all $MgCO_3$ is dissolved and the solution is clear. Then add NaOH until pH is 7.8. When the pH is adjusted to 7.8, add H_2O until a final volume of 50 mL. Store at room temperature.
3. $MgSO_4$ (50 mM, 50 mL). Add 616 mg of $(MgCO_3)_4\cdots Mg(OH)_2\cdots 5H_2O$ (magnesium sulfate heptahydrate) to 50 mL H_2O in a glass bottle or beaker and stir until dissolved. Store at room temperature.

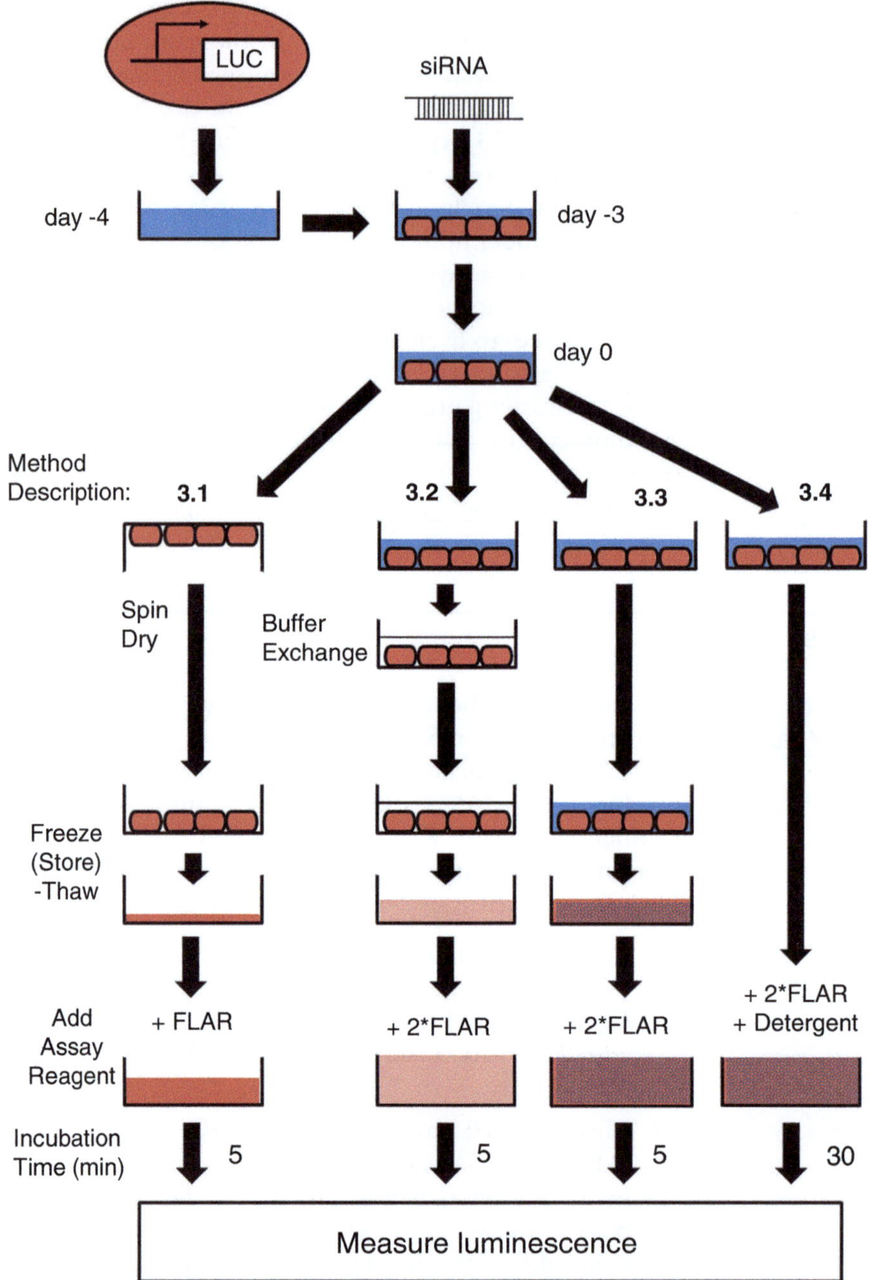

Fig. 1 Schematic depiction of alternative workflows for high-throughput Firefly luciferase reporter assays. The four methods, described in Subheadings 3.1–3.4, were validated in 96-well format using reporter cells carrying a p53-dependent luciferase gene. Luciferase expression was modulated using siRNA silencing TP53; siRNAs silencing p53-inhibitor PPM1D or SYVN1; or irrelevant control nontargeting siRNA. Cells were seeded 4 days and transfected with siRNAs 3 days before start of the luciferase activity assay. Subheadings 3.1–3.3 include a freeze-thaw step. This allows optional (long-term) storage until readout. Subheading 3.1 yields the highest luminescence values; Subheading 3.2 approximately 30% reduced values; and Subheadings 3.3 and 3.4 approximately 70–80% reduced values [8]. Using Subheading 3.1, modulation of p53 activity by siRNA was reproducibly detected in the HTS setting, with siTP53 reducing luminescence five- to tenfold and siPPM1D and siSYVN1 increasing luminescence twofold in different cell lines [8]

4. EDTA (10 mM, 50 mL). Add 146.1 mg EDTA (ethylenediamine-tetraacetic acid) to 50 mL H_2O in a glass bottle or beaker and stir until dissolved. Store at room temperature.
5. DTT (500 mM, 1.5 mL). Add 115.6 mg DTT (dithiothreitol) to a 2 mL Eppendorf tube. Add 1.5 mL H_2O to the tube and vortex until dissolved. Store at −20 °C for a maximum of 3 weeks.
6. ATP (25 mM, 1 mL). Add 13.8 mg ATP (adenosine triphosphate) to an Eppendorf tube. Add 1 mL H_2O to the tube and vortex until dissolved. Store at −20 °C for a maximum of 3 months.
7. D-luciferin (25 mM, 1 mL). Add 7.02 mg D-luciferin (benzothiazole) to an Eppendorf tube. Add 1 mL H_2O to the tube and vortex until dissolved. Store at −20 °C for a maximum of 3 months.
8. Triton X-100, Antifoam 204 and Prionex are purchased as a 100% liquid and need no further preparation.

2.2 Preparation of Assay Reagents

Assay reagents are freshly prepared for each experiment and are kept at room temperature until use on the same day. *See* Subheading 3 for the amounts needed. The final concentrations of the assay reagent components during measurement in all four alternative methods are 20 mM Tricine, 1.07 mM $MgCO_3$, 2.67 mM $MgSO_4$, 100 μM EDTA, 20 mM DTT, 125 μM ATP, and 100 μM D-luciferin. Optional additions are 1% (v/v) Triton X-100, 0.6% (v/v) Prionex, and 0.1% (v/v) Antifoam 204.

1. Rinse buffer (for use in the medium-for-buffer exchange method). Per 1 mL rinse buffer, mix 100 μL Tricine stock, 107 μL $MgCO_3$ stock, 53 μL $MgSO_4$ stock, 100 μL EDTA stock, 40 μL DTT stock, and 600 μL H_2O.
2. Firefly luciferase assay reagent (FLAR; for use in the dry well method). Per 1 mL FLAR, mix 100 μL Tricine stock, 107 μL $MgCO_3$ stock, 53 μL $MgSO_4$ stock, 100 μL EDTA stock, 40 μL DTT stock, 5 μL ATP stock, 4 μL D-luciferin stock, and 591 μL H_2O.
3. Double FLAR (for use in the homogeneous method and the medium-for-buffer exchange method with mechanical cell disruption). Per 1 mL Double FLAR, mix 200 μL Tricine stock, 210 μL $MgCO_3$ stock, 110 μL $MgSO_4$ stock, 200 μL EDTA stock, 80 μL DTT stock, 10 μL ATP stock, 8 μL D-luciferin stock, and 182 μL H_2O.
4. Double FLAR with Detergent (for use in the homogeneous method with detergent cell disruption). Per 1 mL Double FLAR with Detergent, mix 200 μL Tricine stock, 210 μL

$MgCO_3$ stock, 110 µL $MgSO_4$ stock, 200 µL EDTA stock, 80 µL DTT stock, 10 µL ATP stock, 8 µL D-luciferin stock, 10 µL Triton X-100, and 172 µL H_2O (*see* **Note 6**).

2.3 Preparation of Assay Plates with Reporter Cells

1. Seed and culture cells containing the luciferase reporter construct in white-walled 96-wells culture plates (*see* **Note 7**) and conduct the experiment as designed (*see* **Note 8**). A preferred culture volume is 100 µL per well.
2. Take precautions against evaporation during cell culture, to limit differences in volume between wells. This is especially important when using the homogenous assay (*see* **Note 9**).
3. Always include negative controls of untreated reporter cells at least in triplicate on each assay plate. These controls serve to correct for plate differences.
4. Preferably, positive controls that show upregulation and/or downregulation of luciferase activity are also included.
5. Optionally, a separate quality control plate with untreated reporter cells and, if available, positive control treated reporter cells is taken along in the experiment. This plate is used to check the quality of freshly prepared assay reagent before assay plates are analyzed.

3 Methods

All HTS methods (*see* **Note 10**) start with the reporter cells cultured in 96-wells culture plates (Subheading 2.3). The methods describe the preparation of the plates for luminescence readout in a microplate luminometer. For batch wise analysis, the luminescence reader is preferably equipped with a plate handler for automated loading or integrated in a liquid handling robotic platform. We use a Tecan Infinite F200 microplate reader with a Tecan Connect plate loader to process batches of up to 50 plates.

The most elaborate dry well assay yields the highest luminescence values; the medium-for-buffer exchange assay yields slightly lower values. Both variants of the homogeneous assay minimize liquid handling, but yield approximately three- to fivefold lower values compared to the dry well and medium-for-buffer exchange methods.

3.1 Dry Well Assay

1. For each plate, prepare an absorbent paper sheet (*see* **Note 11**) at plate size, i.e., for ANSI/SLAS-standard format plates approximately 12.5 cm × 8.5 cm.
2. Remove most of the culture volume by inverting the plates above a container.

3. Place an absorbent paper sheet on each plate, between the plate and lid.
4. Stack plates upside down in a standard laboratory swing-out centrifuge equipped with microplate rotor holders. Usually, these rotors hold up to four plates per arm (16 plates per run).
5. Centrifuge at 50 × *g* for 1 min. This removes all medium from the wells. The medium is absorbed by the paper sheets.
6. Remove and discard the paper sheets and put the lids back on the plates.
7. Freeze plates at −80 °C for at least 3 h. Optionally, plates can be stored at −80 °C for an extended period until analysis.
8. Thaw plates to room temperature. This usually takes 20–30 min.
9. Prepare fresh FLAR according to Subheading 2.2, **item 2**. For the amount to prepare, take in account the void volume of the dispenser that will be used.
10. Add 30 μL FLAR to each well. In HTS applications, this is usually done using an automated liquid dispenser.
11. Shake the plates for 3 s on an orbital shaker with 1 mm amplitude (*see* **Note 12**).
12. Read luminescence between 5 min and 2 h after adding FLAR.

3.2 Medium-for-Buffer Exchange Assay

1. Gently remove 50% of the culture volume (i.e., usually 50 μL) from each well and replace with the same volume rinse buffer (*see* Subheading 2.2, **item 1**). In HTS applications, this is usually done using an automated cell washer.
2. Repeat **step 1** two more times. The concentration of culture medium is now reduced eightfold.
3. Freeze plates at −80 °C for at least 3 h. Optionally, plates can be stored at −80 °C for an extended period until analysis.
4. Thaw plates to room temperature. This usually takes 20–30 minutes.
5. Prepare fresh Double FLAR according to Subheading 2.2, **item 3**. For the amount to prepare, take in account the void volume of the dispenser that will be used.
6. Add the same volume (i.e., usually 100 μL) Double FLAR to each well. In HTS applications, this is usually done using an automated liquid dispenser.
7. Shake the plates for 3 s on an orbital shaker with 1 mm amplitude (*see* **Note 12**).
8. Read luminescence between 5 min and 2 h after adding Double FLAR.

3.3 Homogeneous Assay with Mechanical Cell Disruption

Follow **steps 3–8** of the medium-for-buffer exchange assay.

3.4 Homogeneous Assay with Detergent Cell Disruption

1. Prepare fresh Double FLAR with Detergent according to Subheading 2.2, **item 4**. For the amount to prepare, take in account the void volume of the dispenser that will be used.
2. Add the same volume (i.e., usually 100 μL) Double FLAR with Detergent to each well. In HTS applications, this is usually done using an automated liquid dispenser.
3. Shake the plates for 60 s on an orbital shaker with 1 mm amplitude.
4. Incubate at room temperature for 30 min to allow complete cell lysis.
5. Read luminescence within 90 min after cell lysis, i.e., between 30 and 120 min after adding Double FLAR with Detergent.

4 Notes

1. Other popular bioluminescent proteins for studies in mammalian cells are Renilla luciferase (from the sea pansy *Renilla reniformis*) [11] and Gaussia luciferase (from the copepod *Gaussia princeps*) [12]. These luciferases catalyse the oxidation of a different substrate (coelanterazine), allowing for dual reporter assays in conjunction with firefly luciferase. In addition, Gaussia luciferase is secreted by mammalian cells, allowing for monitoring reporter gene expression by longitudinal sampling of culture medium.
2. In HTS, large batches of cell culture plates are processed simultaneously. Luminescence readout in laboratory microplate readers including plate transfer usually takes 1–2 min for a 96-well plate. Consequently, considerable time expires between measurement of a first and a last plate in a batch. Unless a plate reader with reagent injector is used, flash-type luminescence decay creates large differences between early and late read values. Although data normalization is a standard step in any type HTS analysis, such large differences negatively affect the quality of a screen. Moreover, if considerable light emission decay occurs within the time required to process a plate, this will result in differences between wells read on the same plate. Such intraplate differences are difficult to correct for in the data analysis procedure. Therefore, prolonged stability of luminescence after reagent addition is highly preferred.
3. Luminescence signal intensity was pH dependent, with higher pH yielding higher peak values. However, at pH above 8.0 we

observed increased light intensity decay. Therefore, we advise to buffer the reaction around pH 7.8. The expensive reagent D-luciferin was saturating at concentrations above 30 μM. The FLAR can thus be made considerably cheaper by limiting the D-luciferin concentration. We use 100 μM in our standard assay, but lower concentrations are allowed. Luminescence signal intensity was also ATP and DTT concentration dependent. At high DTT concentration and at high or low ATP concentration, however, increased signal decay was observed. Therefore, recommended concentrations are: 20 mM DTT and 125 μM ATP. CoA is best omitted from the FLAR. While 250 nM CoA boosted the flash luminescence by four- to fivefold, it strongly reduced luminescence stability. With the recommended FLAR composition, light emission decay was <1%/min for at least 2 h after initiation of the reaction [8].

4. In the homogeneous assay, slightly higher luminescence values are reached when lysis is induced by addition of 1% Triton X-100 than by freeze-thawing the cells. However, due to delayed Triton X-100 induced lysis it took 30 min to reach maximal light emission [8]. Hence, if Triton X-100 detergent is used to lyse cells, luminescence readout should start after 30 min incubation. In contrast, upon mechanical disruption of the cells luminescence can be read immediately.

5. It is not necessary to remove all culture medium. In our hands, reducing the medium concentration eightfold, i.e., three 50% medium-for-buffer exchanges, yielded similar results as complete medium removal [8]. In HTS applications, this can be done with a laboratory plate washer. In particular for cell lines with limited adherence to the culture plate, a gentler cell washer is preferred. We use a Molecular Devices AquaMax 4000 with Cell Wash Head for this purpose. If this equipment is not at hand, the centrifugation method described here can be used. It uses a standard laboratory swing-out centrifuge with microplate rotor holders. We use a Hettich Rotanta 460R centrifuge for this purpose. This method removes all medium and yields similar results as a method in which all culture medium is carefully aspirated manually [8]. Clearly, in HTS applications centrifugation is preferred over manual aspiration.

6. Optionally, Double FLAR with Detergent can be further supplemented with 6 μL Prionex and 1 μL Antifoam 204 per mL reagent. This slightly increases luminescence intensity.

7. The methods are described for assays in 96-well format, but can also be used for assays in 384-well format. For this, the same reagent concentrations and assay conditions are used; only volumes need to be reduced.

8. The described protocol is not limited to the use of any particular cell type or Firefly luciferase reporter construct, nor to the type of reporter gene modulation method used. Optimal methods for seeding, culturing and treatment of the reporter cells used should be predefined. Nonadherent cells can only be analyzed using the homogeneous assay. Loosely adherent cells can be analyzed using the homogeneous assay or the assay with gentle medium-for-buffer exchanges. Adherent cells can be analyzed using any of the alternative methods described herein.
9. Use an incubator with a saturated humidified environment. Useful methods to reduce evaporation include placing culture plates in a box also containing a source of water; and excluding the outer rows and columns of a culture plate from the experiment, instead filling these wells with a source of liquid such as culture medium or PBS.
10. Guidance for validation of HTS assay performance characteristics can be found in the NIH Assay Guidance Manual at https://www.ncbi.nlm.nih.gov/books/NBK83783/.
11. We use Whatman GB003 gel blotting paper, but any type clean absorbent paper will do.
12. Usually, plate readers are equipped with an internal plate shaker. In this case, shaking the plates can be done after loading, immediately prior to readout. Alternatively, a separate plate shaking device is used prior to loading the plates into the reader.

Acknowledgment

This work was supported by the Walter Bruckerhoff Stiftung and the Stichting VUmc CCA.

References

1. De Wet J, Wood KV, DeLuca M et al (1987) Firefly luciferase gene: structure and expression in mammalian cells. Mol Cell Biol 7:725–737
2. Gould SJ, Subramani S (1988) Firefly luciferase as a tool in molecular and cell biology. Anal Biochem 175:5–13
3. Joyeux A, Balaguer P, Germain P et al (1997) Engineered cell lines as a tool for monitoring biological activity of hormone analogs. Anal Biochem 249:119–130
4. Welsh S, Kay SA (1997) Reporter gene expression for monitoring gene transfer. Curr Opin Biotechnol 8:617–622
5. O'Brien M, Daily WJ, Hesselberth PE et al (2005) Homogeneous, bioluminescent protease assays: caspase-3 as a model. J Biomol Screen 10:137–148
6. Luker eKE, Smith MCP, Luker GD et al (2004) Kinetics of regulated protein-protein interactions revealed with firefly luciferase complementation imaging in cells and living animals. Proc Natl Acad Sci U S A 101:12288–12293
7. DeLuca M, Wannlund J, McElroy WD (1979) Factors affecting the kinetics of light emission from crude and purified firefly luciferase. Anal Biochem 95:194–198
8. Siebring-van Olst E, Vermeulen C, de Menezes RX et al (2013) Affordable luciferase reporter

assay for cell-based high-throughput screening. J Biomol Screen 18:453–461

9. Fontes R, Dukhovich A, Sillero A et al (1997) Synthesis of dehydroluciferin, Coenzyme A and nucleoside triphosphates on the luminescent reaction. Biochem Biophys Res Commun 237:445–450
10. Fraga H, Fernandes D, Fontes R et al (2005) Coenzyme A affects firefly luciferase luminescence because it acts as a substrate and not as an allosteric effector. FEBS J 272:5206–5216
11. Lorenz WW, McCann RO et al (1991) Isolation and expression of a cDNA encoding *Renilla reniformis* luciferase. Proc Natl Acad Sci U S A 88:4438–4442
12. Tannous BA, Kim D-E, Fernandez JL et al (2005) Codon-optimized Gaussia luciferase cDNA for mammalian gene expression in culture and *in vivo*. Mol Ther 11:435–443

Chapter 3

Using the 2A Protein Coexpression System: Multicistronic 2A Vectors Expressing Gene(s) of Interest and Reporter Proteins

Garry A. Luke and Martin D. Ryan

Abstract

To date, a huge range of different proteins—many with cotranslational and posttranslational subcellular localization signals—have been coexpressed together with various reporter proteins in vitro and in vivo using 2A peptides. The pros and cons of 2A co-expression technology are considered below, followed by a simple example of a "how to" protocol to concatenate multiple genes of interest, together with a reporter gene, into a single gene linked *via* 2As for easy identification or selection of transduced cells.

Key words 2A peptide, CHYSEL, Bicistronic vector

1 Introduction

1.1 Background

Two of the most popular strategies to coexpress multiple genes from one mRNA are the use of internal ribosome entry site (IRES) sequences and 2A oligopeptide sequences (*cis*-acting **hy**drolase **el**ement "CHYSEL"). IRESes are *cis*-acting sequences able to sequester translation initiation factors and subsequently ribosomal subunits to initiate translation without a 5′ cap structure [1]. Bicistronic and multicistronic vectors containing IRES elements have been widely used for in vitro and in vivo applications such as coexpression of selective markers or reporter genes [2, 3]. Since their discovery within the RNA genomes of picornaviruses [4], similar structures have been identified in many other viral RNAs and in a variety of cellular mRNAs [5]. Currently, the most commonly used IRESes originate from picornaviruses such as encephalomyocarditis virus (EMCV) and foot-and-mouth disease virus (FMDV) [6, 7]. The benefits of IRES-mediated expression strategies include: ensured coexpression of genes; ability to control the relative expression of different genes by varying the strength of the IRES applied to each gene; possibility of adding subcellular

Robert Damoiseaux and Samuel Hasson (eds.), *Reporter Gene Assays: Methods and Protocols*, Methods in Molecular Biology, vol. 1755, https://doi.org/10.1007/978-1-4939-7724-6_3,

localization sequences to the gene after the IRES. Despite their widespread use, they have two major limitations: they are relatively large (~600 bp), adding to the size of the transgene. Furthermore, in many instances only the first open reading frame (ORF) is expressed strongly (translated in a cap-dependent manner) and IRES-dependent translation of the second ORF (translated in a cap-*in*dependent manner) is much weaker [8–10]. These drawbacks can be overcome by a 2A peptide, a "self-cleaving" peptide first identified in FMDV [11].

In recent years, the 2A-based polyprotein system has been used in a variety of eukaryotic systems, and is considered to be the "go-to" technology for transgene co-expression. The short 2A peptide, 18 amino acids (aa) in length, was first characterized in the central region of the FMDV polyprotein, between the upstream capsid and the downstream replication protein domains. FMDV 2A (abbreviated F2A) comprises two parts, an N-terminal region (without high sequence conservation) and a conserved motif comprising the seven C-terminal residues of 2A and the N-terminal proline of the downstream protein 2B (-D[V/I]xExNPG↓P-; underlined proline comprises the N-terminal residue of 2B). Analysis of recombinant FMDV polyproteins [11] and artificial polyprotein systems in which F2A was inserted between two reporter proteins [12] showed that 2A mediated a ribosomal skipping event in which the synthesis of a specific peptide bond was "skipped": translation terminated at the C-terminus of 2A, but could "reinitiate" at the N-terminal proline of the 2B protein. Although no stop codon is involved, eukaryotic translation release (termination) factors 1 and 3 (eRF1/eRF3) are proposed to release the nascent protein from the ribosome. Our translational model of 2A-mediated cleavage activity is shown in Fig. 1. Known by the various epithets of "ribosome skipping," "StopGo," and "Stop-Carry On" translation, it allows the stoichiometric production of multiple, discrete, products from a single transgene [13, 14]. For simplicity, 2A activity will simply be referred to as "cleavage" below.

1.2 F2A and "2A-Like" Peptide Sequences

Later work shows the "signature" DxExNPGP motif is also present in other virus polyproteins, e.g., insect viruses, dsRNA viruses of humans, insects, and crustaceans [15, 16], active 2A-like sequences in non-LTR retro-transposons (chordates, molluscs, cnidarians, and echinoderms) [17–19] and active 2A-like sequences in genes of the innate immune systems of a wide-range of marine organisms (unpublished data). Of the 2A peptides identified to date, four have been widely used in biomedical research: F2A; "E2A" from equine rhinitis virus (ERAV), "P2A" from porcine teschovirus-1 (PTV-1) and "T2A" from *Thosea asigna* virus (TaV) [20–25]. The first three belong to the family *Picornaviridae* and the last is an insect virus [26] (Table 1). Comparing the in vitro cleavage of these different 2As inserted between green fluorescent protein (GFP) and

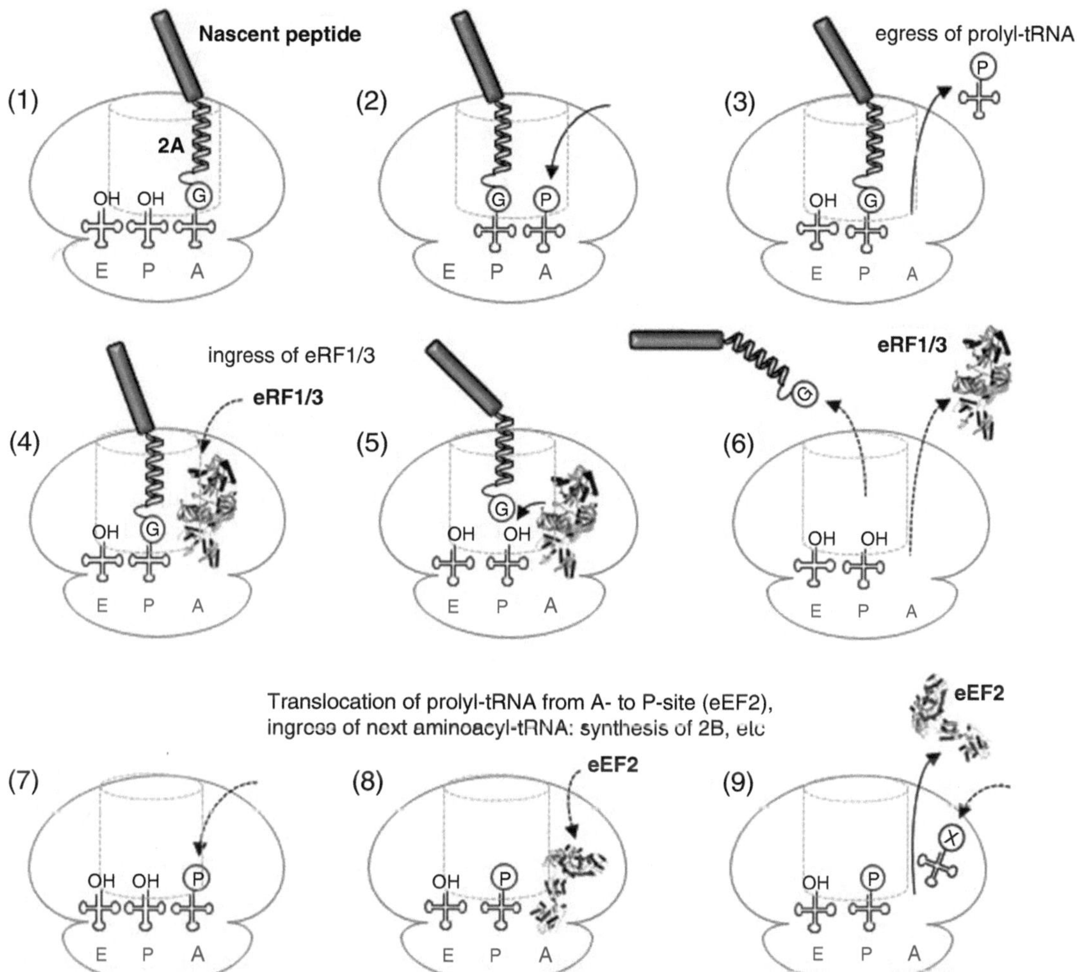

Fig. 1 Schematic representation of the translational model of 2A-mediated "cleavage." The nascent peptide has emerged from the ribosome exit tunnel, whilst 2A is positioned within the exit tunnel (step 1). The nascent peptide-2A and glycyl-tRNA is translocated from the A to P site and prolyl-tRNA enters the A-site (step 2). The model we propose is that the nascent 2A peptide interacts with the exit pore of the ribosome such that the C-terminal portion (-ESNPGP-) is sterically constrained within the peptidyltransferase center (PTC) of the ribosome. This inhibits nucleophilic attack of the ester linkage between 2A-tRNA^{gly} by prolyl-tRNA in the A site—effectively stalling, or pausing translation. The failure to generate a new peptide bond leads to dissociation of prolyl-tRNA from the A-site of the ribosome (step 3), required for entry of RF into the A-site (step 4). It has been shown that this block is relieved by the action of translation release factors eRF1 and eRF3, hydrolyzing the ester linkage (step 5) and releasing the nascent protein (step 6). eRF1 leaves the complex, eRF3 being involved in this process (step 6). Two, mutually exclusive, outcomes may then arise: (a) translation terminates at the C-terminus of 2A, or, (b) prolyl-tRNA (re)enters the A-site (step 7), is translocated by eukaryotic elongation factor 2 (eEF2) from the A- to the P-site (step 8) allowing the next amino-acyl tRNA to enter the A-site to synthesize the downstream sequences (step 9)

Table 1
2A and "2A-like" sequences most commonly used for protein coexpression

Source	Abbreviation	2A/2A-like sequence	Number of studies[a]	Recoding Activity
Foot-and-mouth disease virus	$F2A_{18}$	-LNFDLLKLAG**DVESNPG P**-	223	***
	$F2A_{30}$	-HKQKIVAPVKQLNFDLLKLAG**DVESNPG P**-		****
Equine rhinitis A virus	E2A	-QCTNYALLKLAG**DVESNPG P**-	24	****
Porcine teschovirus 1	P2A	-ATNFSLLKQAG**DVEENPG P**-	76	****
Thosea asigna virus	T2A	-EGRGSLLTCG**DVEENPG P**-	121	*****

The DxExNPGP motif conserved among 2A/2A-like sequences is shown in bold, red, typeface
[a]Number of published papers citing the use of each type of 2A-like sequence

β-glucuronidase (GUS), we have shown that $T2A_{20}$ has the highest cleavage efficiency followed by $E2A_{20}$, $P2A_{20}$ and $F2A_{20}$ [15]. Using the T-cell receptor (TCR):CD3 complex as a test system, Szymczak and colleagues demonstrated that $F2A_{22}$ and $T2A_{18}$ have higher efficiency (~100%) than $E2A_{20}$ [21]. In three commonly used human cell lines, zebrafish embryos and mouse liver, cleavage and targeting of NLS-EGFP and mCherry-CAAX to the nucleus and plasma membrane, respectively, was the most efficient using the $P2A_{19}$ plasmid, followed by $T2A_{18}$, $E2A_{20}$, and $F2A_{22}$ [27]. The availability of a range of 2A-like sequences can be useful when the presence of direct repeats within a construct might cause problems. To minimize the risk of homologous recombination and genetic instability, it is wise to use different 2A peptide sequences if more than two genes are being linked.

1.3 Polycistronic Reporter Systems

2A peptide cleavage has been studied using various artificial reporter polyprotein systems comprising chloramphenicol acetyltransferase (CAT), GUS, and fluorescent proteins (FPs, e.g., GFP, RFP, YFP) in various cell types, as well as targeting to various subcellular localizations in plants [28, 29] and animal cells [22, 24, 30]. To assess visually 2A cleavage in transgenic mice, Trichas and coworkers linked a membrane-localized TdTomato gene [*Myr-tdTomato*] to a nuclear localized enhanced GFP gene [*H2B-eGFP*] using the T2A sequence [24]. Mutually exclusive localization of TdTomato and eGFP to the membrane and nucleus, respectively, was observed in both cultured cells and cell-lineages within the transgenic mice, demonstrating complete and reliable 2A-mediated processing. Moreover, targeted expression was apparent in all tissues examined throughout development and adulthood and function remained constant across several generations. By the same token, 2A can also be used to monitor the expression of transgenes. Mucopolysaccharidose type 1 (MPS-1; Hurler syndrome) is a lysosomal monogenic disorder caused by mutations in the gene for α-L-iduronidase (IDUA)—to alleviate progression of the disease MPS-1 patients need a constant supply of IDUA. To achieve stable gene expression in mice, P2A and T2A were used to construct a tricistronic gene therapy vector bearing the human *IDUA* gene along with the firefly luciferase and DsRed2 reporter genes [*IDUA*-P2A-luciferase-T2A-DsRed] [31]. Efficient cleavage was observed and all three proteins were functional in vitro and in vivo, allowing for therapeutic enzyme levels that could be tracked by DsRed2 expression in cells and luciferase imaging in whole animals.

Genome editing (GE) technologies based on custom-designed nucleases such as zinc finger nucleases (ZFNs), transcription activator-like effector nucleases (TALENs) and RNA-guided nucleases (RGNs) [32–34] are opening up unprecedented possibilities for targeted changes within the genome of cells and organisms. The double-strand break (DSB) made by these targetable nucleases stimulates the cell's natural DNA-repair mechanisms, typically leading to one of two outcomes: nonhomologous end joining (NHEJ) to introduce insertions/deletions, or if an appropriate target-homologous donor is supplied, homology directed repair (HDR) [35]. ZFNs and TALENs, although they share the same *Fok*I-derived nuclease domain, differ in that they use distinctive DNA-binding arrays: ZFNs use zinc finger arrays [36] and TALENs use TAL effector repeat arrays [37]. The clustered regularly interspaced short palindromic repeat (CRISPR)/Cas RGNs rely on base pairing between a "guide" RNA and the DNA target for recognition and a Cas nuclease for DNA cleavage [32, 38]. While the nuclease triad has shown to facilitate GE, each approach has its pros and cons, which may dictate the choice for a given experiment [34, 39, 40].

One of the major bottlenecks to the application of programmable nucleases is the lack of systems to select or enrich genome-modified cells. GE by using 2A peptide coupled co-expression of nuclease and fluorescent proteins combined with fluorescence-activated cell sorting (FACS) can aid selective enrichment of transfected cells [41–45]. For ZFNs, this general strategy, together with delivery of nucleases as mRNA and donors as single-stranded oligodeoxynucleotides (ssODNs) provided a proof-of-concept for high knock-in efficiencies in several diverse applications. Importantly, observed off-target effects were minimal [45]. Because ZFNs can bind with some degree of mismatching, cutting at off-target sites in the genome can occur, which may be minimized by using lower ZFN concentrations and optimizing binding to the on-target site [46, 47]. Since reporter signal in the 2A approach is a direct measure (1:1) of nuclease levels, it may be easier to balance nuclease levels for an optimal on/off target ratio using this method. In relation to other nucleases used for GE, this method also works well for TALENS [41] and CRISPR/Cas9 paired nickases

[45, 48]. To expand the experimental toolkit, antibiotic (hygromycin) selection and magnetic H-2K^k cell selection methods have been developed to enrich transfected cells with ZFN/TALEN-induced mutations [49]. These reporters contain the target sequences of the engineered nucleases and express Hygro^R and H-2K^k, respectively, only when insertions or deletions are generated in the target cell sequences by the activity of engineered nucleases. Further, because both reporters express GFP in addition to Hygro^R and H-2K^k (2A-Hygro^R-eGFP and eGFP-2A-H-2K^k) they can also be used for flow cytometric enrichment of edited cells. Additionally, so called "gene trap" approaches use a 2A peptide sequence that links the reading frame of the gene of interest with the expression of a selectable marker [44, 50].

Fluorescent reporter systems have also proven indispensable for evaluating the two major DNA repair pathways [51]. The two-color "Traffic Light Reporter" (TLR; *eGFP-T2A-mCherry*) developed by Scharenberg and colleagues uses a flow cytometric assay to simultaneously detect both gene repair and mutagenic NHEJ at a single targeted site [52]. HDR using a donor template carrying an intact *eGFP* results in green fluorescent cells. If mutagenic NHEJ repairs the break, a shift in the reading frame of the *eGFP* reporter can result in translation of the *T2A-mCherry*, leading to red fluorescent cells. In an independent study Kühn et al. adopted three different strategies (gene silencing, small-molecule inhibition or proteolytic degradation) to abolish NHEJ in a "traffic light," leading to enhanced HDR for CRISPR-Cas9-induced gene targeting [53].

1.4 Vector Design and Construction Considerations

This technology has been used for human gene therapies targeting cancer, production of induced human pluripotent stem cells for regenerative medicine, creation of transgenic animals and plants with improved nutritional properties, and the production of high-value proteins for the pharmaceutical industry [54–56]. The potential of this system is vividly demonstrated by the expression of two functional biochemical pathways encoded by nine genes converted into a single polycistronic T2A peptide based transcript in the yeast *P. pastoris*: the highest number of genes expressed in such a co-ordinated fashion so far [57]. In order to avoid sequence identities the T2A coding sequence was employed in different nucleotide sequence versions. In this chapter we offer our perspective on the use of 2A peptide sequences and how we think some "problems" might be addressed along with some examples from other laboratories where similar issues have been tackled. This is followed by an illustrative "how to" example that describes the generation of a 2A peptide-based vector that directs coexpression of the therapeutic human (*IDUA*) gene along with the GFP reporter gene.

Our analyses of this translational "recoding" event are based upon the creation of an artificial polyprotein system comprising

GFP (stop codon removed) linked, *via* the F2A sequence, to GUS (initiation codon removed) to create a single ORF (plasmid [*GFP-F2A-GUS*] [16]. Significantly, this version has been used widely in other laboratories and is available on request without restriction of use. Mutagenesis experiments showed that F2A cleavage is either reduced or ablated by amino acid substitution over most of its length. In general, mutations of conserved amino acids have more pronounced effects than changes to nonconserved ones [15, 58, 59]. Sequences immediately upstream of F2A are known to be either critical or very important for cleavage: N-terminal extension of F2A by 5aa of 1D improved "cleavage," but extension by 14aa of 1D or longer (21 and 39aa) produced complete "cleavage" and an equal stoichiometry of GFP and GUS [15]. In a follow-on study, optimal and suboptimal lengths of F2A were identified both in vitro and in vivo using F2A of various lengths to link GFP and CherryFP reporter proteins [60]. After "tweaking," we recommend researchers opt for $F2A_{30}$ (+11aa1D)—this 2A proved to be the most favourable in terms of both length and cleavage efficiency and was unaffected by the sequence of the upstream gene [60, 61]. The efficiency of processing has also been increased by insertion of various spacer sequences such as a V5 epitope tag: -GKPUPNPLLGLDST- [62], a 3×FLAG epitope tag: -DYKDHDG-DYKDHDI-DYKDDDDK- [63], or a glycine-serine linker: -GSG- or –SGSG- [21, 22, 64, 65] between the upstream protein and the 2A sequence. This provides more flexibility between the two proteins making the inhibitory residues more distal. Finally, gene order effects on the cleavage efficiency, stability, and localization must be considered—proteins that require authentic termini, or are N-/C-terminally modified, can be introduced as the first or final polyprotein domain, respectively [66–69].

The study of 2A and its extensive use in biotechnological and biomedical applications shows that these sequences work in all eukaryotic cell-types tested to date: plant [20], fungi [70], yeast [71], insect [72], and mammalian [12], but not in bacteria. A potential drawback of this system is the possibility that the residual 2A peptide at the end of the N-terminal protein may affect protein trafficking or function, or contribute to the antigenicity of the proteins [10, 67]. You can in any case counter this to some extent by using shorter versions of 2A or placing genes that encode proteins amenable to C-terminus modifications upstream of the 2A sequence. In the case of proteins that are targeted through the endoplasmic reticulum (ER), cell surface or exocytic pathways, the 2A extension can be "trimmed" away by the inclusion of a furin cleavage site (-↓RRRR-, -↓RKRR-, -↓RRKR-) between the upstream protein and 2A [23, 69]. Furin is a cellular endoprotease localized on the trans-Golgi networks of virtually all cell types [73]. Another potential solution to this problem, at least for membrane-associated proteins, is to incorporate a glycosylphosphatidylinositol (GPI) anchor signal sequence upstream of 2A. This signal peptide is post-translationally replaced with a GPI anchor in the ER and should therefore result in the complete removal of any C-terminal extensions [74]. In plants, the first nine amino acids of the LP4 peptide of *Impatiens balsamina* (-SN↓AADEVAT-) was connected to the 20aa F2A (-QLLNFDLLKLAGDVESNPG↓P-) to generate a similar hybrid linker peptide (LP4/2) [75, 76]. The "unwanted" tag may remain—antibodies have been raised against the consensus 2A sequence, which thus serves as a useful target for identifying proteins in biochemical assays [30, 71]. Lastly, it should be noted that all proteins produced after the 2A peptide will have an amino-terminal proline. Although this addition confers a long half-life (>20 h) on proteins [77], it might also interfere with the function of a given protein, and this will have to be determined empirically for each application.

Everything considered, we now describe the production of a basic vector using the T2A sequence (short version) to link the expression of a human *IDUA* gene to a reporter gene (Fig. 2). *Indirect* imaging of therapeutic transgene expression requires proportional and constant coexpression of both genes over a wide range of transgene expression levels. Since 2A function is primarily dependent on eukaryotic ribosomes [12], it can be presumed that translation of the linked proteins is independent of the cell type and differentiation status. This is a clear advantage over other genetic strategies (alternative splicing, internal promoters, or IRES).

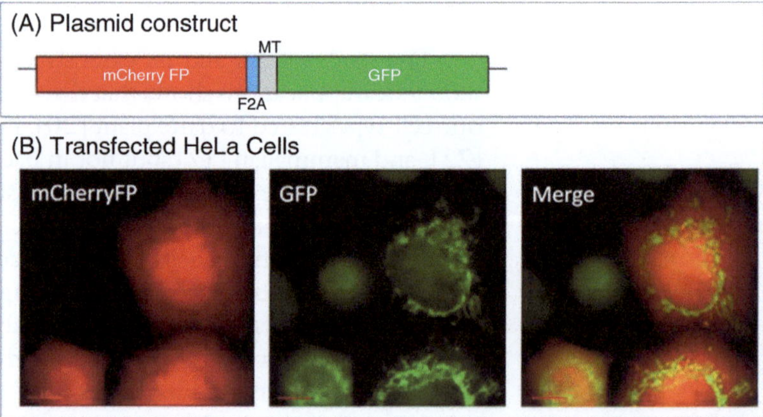

Fig. 2 Use of 2A to coexpress mCherryFP and GFP targeted to mitochondria. The plasmid encoding pmCherryFP-F2A-mtGFP (Panel **A**) was used to transfect HeLa cells (Panel **B**). Lacking any form of targeting signal, the mCherryFP is uniformly distributed throughout the cytoplasm and diffuses into the nucleus. The mitochondrial targeting signal (MT) fused to GFP localizes this fluorescent protein to the mitochondria. Scale bars = 10 μm

2 Materials

The following reagents can be obtained from commercial suppliers: GoTaq® Flexi DNA Polymerase, T4 DNA Ligase, PCR Nucleotide Mix, restriction enzymes, Ethidium bromide, pGEM®-T Easy Vector System, Wizard® SV Gel and PCR Clean-Up System, JM109 High Efficiency Competent Cells and Quick *TnT* rabbit reticulocyte lysate kit (Promega, Madison, WI, USA); human α-L-iduronidase (IDUA) (OriGene, Rockville, MD, USA); rabbit anti-FMDV 2A antiserum raised against the synthetic peptide NH_2-LLNFDLLKLAGDVESNPGP- COOH (Dundee Cell Products, Dundee, UK); Luria-Bertani (LB) Medium (ForMedium™, Norfolk, UK); oligonucleotides (IDT, Leuven, Belgium); Plasmid DNA Mini Kit II (VWR International Ltd., Leighton Buzzard, UK); ^{35}S methionine (MP Biomedicals, Santa Ana, CA, USA); Ampicillin Sodium Salt, IPTG (isopropyl-β-D-thiogalactopyranoside) and X-Gal (5-bromo-4-chloro-3-indolyl-β-D-galactoside) (Melford Laboratories Ltd., Ipswich, UK). The reporter [*GUS-F2A-GFP*] plasmid is available on request without restriction of use.

3 Methods

3.1 Primer Design (See Notes 1 and 2)

The polymerase chain reaction (PCR) strategy, shown in Fig. 3, has the advantage that restriction enzyme (RE) sites are created at desired positions within the "forward" and "reverse" oligonucleotide PCR primers used for cloning. In the example shown in Fig. 3, a *BamHI* site has been created in the forward primer immediately upstream of the initiation codon of the *IDUA* gene (Fig. 3, panels B, **D**: *IDUA* sequences shaded in yellow). To fuse the T2A sequence (shaded in grey) onto the *IDUA* gene and maintain a single open reading frame, the reverse primer design removes the *IDUA* stop codon: the cost of long primer sequence synthesis (100 bp and above) makes it possible to use the reverse primer to encode the entire T2A sequence—together with 24 bases complementary to the 3′-end of *IDUA* for annealing to the template *IDUA* gene (Fig. 3, panels **B**, **D**). An *ApaI* site is also incorporated into this reverse primer (Fig. 3, panels **C**, **D**) such that the PCR product produced will form a gene "block" that may easily be inserted into the final "vector" (Fig. 3, panel **A**).

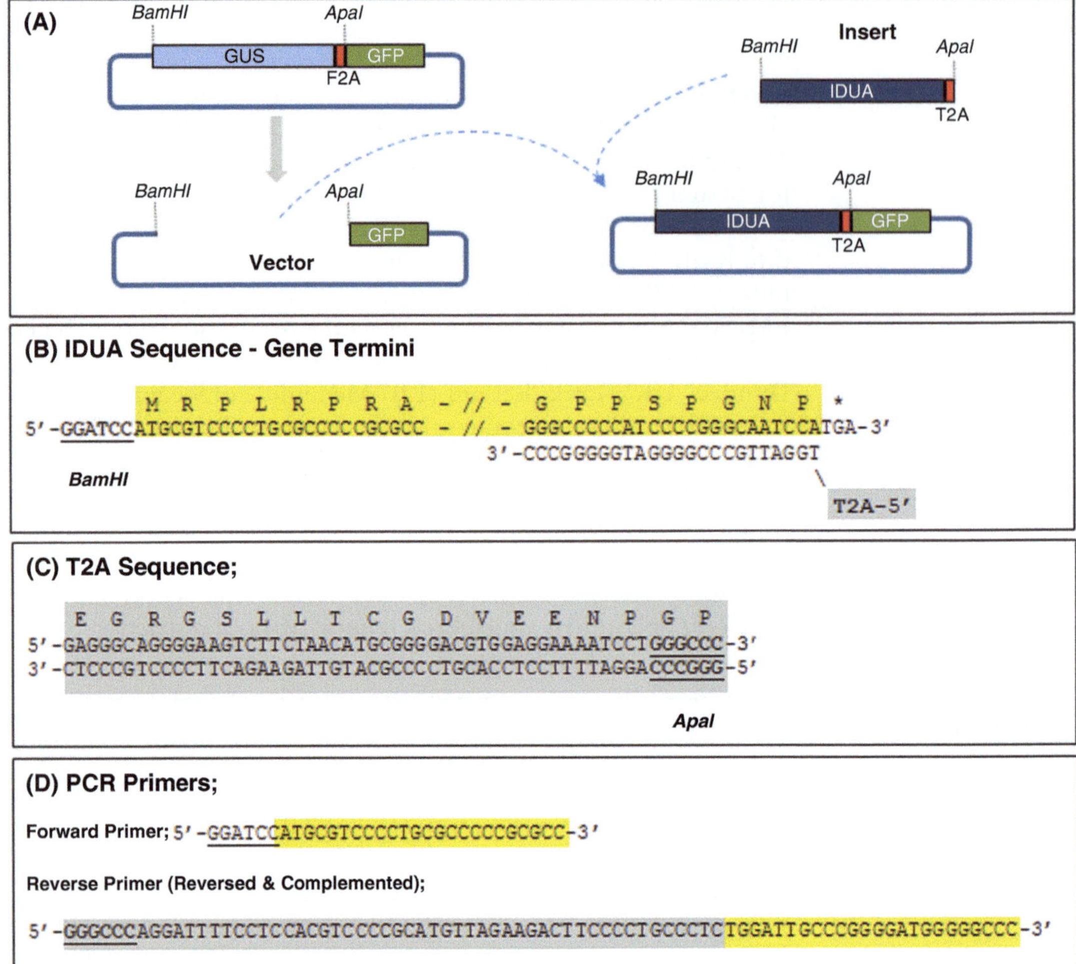

Fig. 3 Schematic diagram showing the primer design strategy. The IDUA-T2A amplified product (2019bp) restricted with *BamH1* and *Apa1* and cloned in-frame into vector pGUS-F2A-GFP digested with the same enzymes produces pIUDA-T2A-GFP (Panel **A**). Sequence of the termini of the human α-L-iduronidase mRNA (IDUA; transcript variant 1; NCBI Reference Sequence: NM_000203.4: *IDUA* sequences shaded in yellow). The forward primer also encodes a *BamHI* restriction enzyme site which will become part of the PCR product and will be used in subsequent cloning steps. The reverse primer is complementary to the 3′ region, omits the stop codon, but also encodes the T2A sequence (shaded in grey)—this will also become part of the PCR product (Panel **B**). Amino acid and nucleotide sequence of the T2A peptide used in this study, taken from constructs described by Osborn et al. [31]. The *ApaI* site at the C-terminus of the T2A sequence is used for subsequent cloning steps (Panel **C**). The forward and reverse primers used to amplify the *IDUA* gene plus T2A together with the restriction sites used to clone the insert (Panel **D**)

3.2 Amplification PCR of IDUA cDNA (See Note 3)

Component	Final volume (μL)	Final concentration
GoTaq® Flexi Buffer	10	1×
$MgCl_2$ Solution, 25 mM	2	1.0 mM
PCR Nucleotide Mix, 10 mM each	1	0.2 mM each dNTP
Upstream primer	1	1 μM
Downstream primer	1	1 μM
GoTaq® DNA Polymerase (5 u/μ L)	0.5	2.5 u
IDUA cDNA template	1	~0.5 μg/50 μL
Nuclease-Free Water to	50 μL	

Thermal Cycling Conditions for PCR Amplification of *IDUA* cDNA

Step	Temperature (°C)	Time (min)	Number of cycles
Initial denaturation	94	2.0	1
Denaturation	94	0.5	
Annealing	55	0.5	25[a]
Extension	71	1.0	
Final extension	71	7.0	1
Soak	4	∞	1

[a]The denaturation / annealing / extension steps are performed sequentially, a total of 25 times

An aliquot of the PCR reaction should be analyzed on an agarose gel before use in the pGEM®-T Easy vector ligation reaction to verify product purity. Gel extraction prior to ligation is recommended to remove any remaining primers and other unwanted background bands. The Biomath calculator (www.promega.com/biomath) can be used to calculate the appropriate amount of PCR product (insert) to include in the ligation reaction described below. Generally, incubation overnight at 4 °C will produce the maximum number of transformants.

3.3 Ligation Protocol (See Note 4)

Reaction component	Standard reaction (μL)	Positive control (μL)	Background control (μL)
2× Rapid Ligation Buffer, T4 DNA Ligase	5	5	5
pGEM®-T Easy Vector (50 ng)	1	1	1
PCR product[a]	3	–	–
Control Insert DNA	–	2	–
T4 DNA Ligase (3 Weuss units/μL)	1	1	1
Nuclease-free water to a final volume of	10	10	10

[a]Molar ratio of PCR product:vector may require optimization

3.4 Transformation Protocol Using the pGEM®-T Easy Vector Ligation Reactions

1. Successful cloning of an insert into the pGEM®-T Easy Vector interrupts the coding sequence of β-galactosidase; recombinant clones can be identified by blue/white screening on LB plates with ampicillin/IPTG/X-Gal. Prepare plates for the standard reaction and experimental controls.
2. Briefly centrifuge the tubes containing the ligation reactions. Add 2 μL of each reaction to a 1.5 mL microcentrifuge tube on ice.
3. Place the High Efficiency Competent *E. coli* Cells (strain JM109) in an ice bath until thawed (~5 min). Mix the cells by gently flicking the tube.
4. Carefully transfer 50 μL of cells to the microcentrifuge tubes from **step 2**. Gently flick the tubes and incubate on ice for 20 min.
5. Heat-shock the cells for 50 s in a water bath at 42 °C. Return the tubes to ice for 2 min.
6. Add 950 μL room temperature LB to the tubes containing cells transformed with ligation reactions and incubate for 1.5 h at 37 °C with shaking (~150 rpm).
7. Plate 100 μL of each transformation culture onto LB/ampicillin/IPTG/X-Gal plates. For a higher number of colonies, the cells may be pelleted by centrifugation at 1000 × *g* for 10 min, and resuspended in 100 μL of LB medium.
8. Incubate plates overnight at 37 °C.

3.5 Plasmid Construct

Isolate a single white colony, and inoculate a culture of 10 mL LB medium containing 100 μg/mL ampicillin. Incubate for 12–16 h at 37 °C with vigorous shaking (~300 rpm). The Plasmid DNA

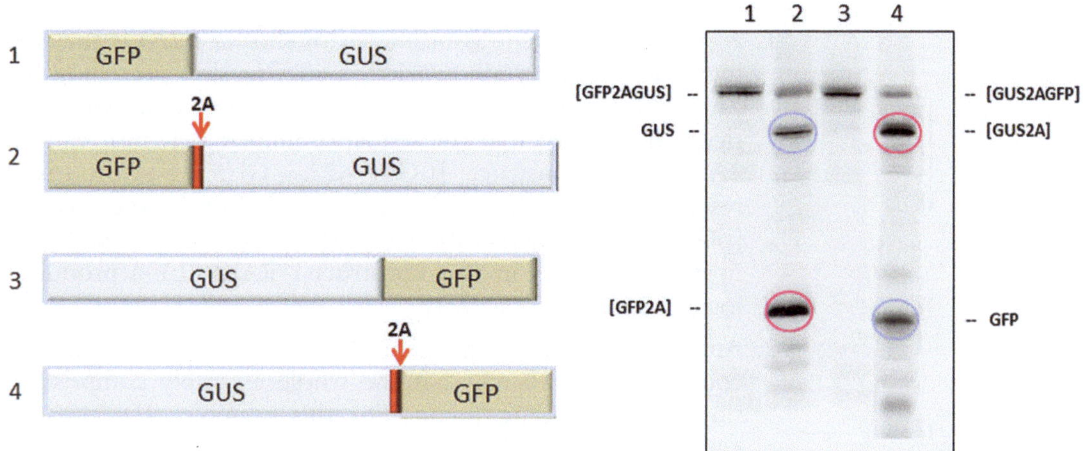

Fig. 4 Analysis of 2A-mediated cleavage. Artificial reporter polyproteins (boxed areas) used to programme in vitro translation systems are shown together with translation profiles obtained from rabbit reticulocyte lysates (right). Control constructs pGFPGUS and pGUSGFP produce only a single translation product—the [GFPGUS] and [GUSGFP] fusion proteins respectively. The translation profile from the pGFP2AGUS construct shows three major products: uncleaved [GFP2AGUS], and the cleavage products [GFP2A] and [GUS]. The profile pGUS2AGFP also shows three major products: uncleaved [GUS2AGFP], and the cleavage products [GUS2A] and [GFP]. The cleavage products upstream of 2A are highlighted in red and the downstream products shown in blue

Mini Kit II is used to isolate plasmid DNA from cultures. A single digest with *BstZI*, *EcoRI* or *NotI* will release inserts and identify clones that contain the PCR product of interest. The inserts can then be sequenced using the T7 promoter primer or SP6 promoter primer. The amplified product restricted with *BamH1* and *Apa1* and cloned into our pcDNA™ 3.1 mammalian expression vector [pGUS-F2A-GFP] [13] digested with the same enzymes produces [pIUDA-T2A-GFP].

3.6 Transcription and Translation Reactions (See Note 5)

We have effectively used in vitro coupled transcription/translation cell free systems to assess 2A cleavage [13, 15–17] (Fig. 4). Briefly, plasmid DNA (100 ng) is used to program the lysate master mix (10 μL) supplemented with ^{35}S-methionine (10 μCi). Reactions are incubated at 30 °C for 90 min before the addition of 2× SDS-PAGE loading buffer. Translation reactions (5 μL aliquots) are analyzed by SDS-PAGE (10%) and autoradiography.

4 Notes

1. A potential problem associated with shorter F2As is that the introduction of certain restriction enzyme sites preceding the 2A peptide sequence may affect cleavage [65]. Our own experience has shown that the sequence immediately upstream of 2A in the construct [pGFP-F2A$_{20}$-GUS] (-SGSRGAC-) resulted in a highly active F2A [15], but in the case of [pGFP-F2A$_{20}$-CherryFP] [61] the sequence immediately upstream of F2A in this construct (-RAKRSLE-) produced lower F2A cleavage activity.
2. Any gene sequences upstream of the 2A must have their stop codons removed such that the whole assembly comprises a single open reading frame. For this example, *IDUA* (stop codon removed) linked, *via* the T2A sequence, to GFP (initiation codon removed) to create a single ORF ([IDUA-T2A-GFP].
3. The pGEM®-T Easy vector has been linearized and a single thymidine added to both 3′-ends. Thermostable DNA polymerases with proofreading activity, such as *Pfu* DNA Polymerase, generate blunt-ended fragments. PCR products generated using this polymerase can be modified using an A-tailing procedure and ligated into the pGEM®-T Easy Vector [78]. Using this method, only one insert will be ligated into the vector, as opposed to multiple insertions that can occur with blunt-ended cloning.
4. We recommend using Experimental Controls with the ligation. The supplied positive control will allow you to determine whether the ligation is proceeding efficiently—typically >60% of the colonies should be white. The background control allows determination of the number of blue colonies resulting from non-T-tailed or undigested vector alone.
5. Depending on the proteins or vector system under study, alternative methods for verifying 2A cleavage may be necessary. An effective way to assess 2A function is to perform transient transfection of HeLa cells (human cervical epithelial carcinoma cells) followed by western blot analysis [30, 60]. Cleavage can be visualized by detection with antibodies against the target proteins or anti-2A serum (antibodies raised against 2A are available on request).

Acknowledgments

The authors gratefully acknowledge the long-term support of our research by the UK Biotechnology and Biological Sciences Research Council (BBSRC) and the Wellcome Trust. The University of St Andrews is a charity registered in Scotland, no. SCO13532.

References

1. Elroy-Stein O, Merrick WC (2007) Translation initiation *via* cellular internal ribosome entry sites. In: Mathews MB et al (eds) Translational control in biology and medicine. Cold Spring Harbor Laboratory Press, Cold Spring Harbor, NY, pp 155–172
2. Martínez-Salas E (1999) Internal ribosome entry site biology and its use in expression vectors. Curr Opin Biotechnol 5:458–464
3. Attal J, Theron MC, Houdebine LM (1999) The optimal use of IRES (internal ribosome entry site) in expression vectors. Genet Anal 15:161–165
4. Pelletier J, Sonenberg N (1988) Internal initiation of translation of eukaryotic mRNA directed by a sequence derived from poliovirus RNA. Nature 334:320–325
5. Minskaia E, Luke GA, Ryan MD (2015) Co-expression technologies in eukaryotic cells. In: Zahoorullah S (ed) Textbook of Biotechnology. SM Online Publishers LLC, Austin, TX, pp 1–16
6. Jang SK, Wimmer E (1990) Cap-independent translation of encephalomyocarditis virus RNA: structural elements of the internal ribosome entry site and involvement of a cellular 57-kD RNA-binding protein. Genes Dev 4:1560–1572
7. Belsham GJ (1992) Dual initiation sites of protein synthesis on foot-and-mouth disease virus RNA are selected following internal entry and scanning of ribosomes *in vivo*. EMBO J 11:1105–1110
8. Mizuguchi H, Xu Z, Ishii-Watabe A et al (2000) IRES-dependent second gene expression is significantly lower than cap-dependent first gene expression in a bicistronic vector. Mol Ther 1:376–382
9. Urwin PE, Zubko EI, Atkinson HJ (2002) The biotechnological application and limitation of ires to deliver multiple defence genes to plant pathogens. Physiol Mol Plant Pathol 61 (2):103–108
10. Hasegawa K, Cowan AB, Nakatsuji N et al (2007) Efficient multicistronic expression of a transgene in human embryonic stem cells. Stem Cells 25:1707–1712
11. Ryan MD, King AM, Thomas GP (1991) Cleavage of foot-and-mouth disease virus polyprotein is mediated by residues located within a 19 amino acid sequence. J Gen Virol 72:2727–2732
12. Ryan MD, Drew J (1994) Foot-and-mouth disease virus 2A oligopeptide mediated cleavage of an artificial polyprotein. EMBO J 13:928–933
13. Donnelly MLL, Luke GA, Mehrotra A et al (2001) Analysis of the aphthovirus 2A/2B polyprotein "cleavage" mechanism indicates not a proteolytic reaction, but a novel translational effect: a putative ribosomal "skip". J Gen Virol 82:1013–1025
14. Brown JD, Ryan MD (2010) Ribosome "skipping": "stop-carry on" or "stopgo" translation. In: Atkins JF, Gesteland RF (eds) Recoding: expansion of decoding rules enriches gene expression. Springer, New York, NY, pp 101–122
15. Donnelly MLL, Hughes LE, Luke GA et al (2001) The "cleavage" activities of FMDV 2A site-directed mutants and naturally occurring "2A-like" sequences. J Gen Virol 82:1027–1041
16. Luke GA, de Felipe P, Lukashev A et al (2008) The occurrence, function and evolutionary origins of "2A-like" sequences in virus genomes. J Gen Virol 89:1036–1042
17. Odon V, Luke GA, Roulston C et al (2013) APE-type non-LTR retrotransposons of multicellular organisms encode virus-like 2A oligopeptide sequences, which mediate translational recoding during protein synthesis. Mol Biol Evol 30:1955–1965
18. Luke GA, Roulston C, Odon V et al (2014) Lost in translation: the biogenesis of non-LTR retrotransposon proteins. Mob Gen Elem 3(6): e27525
19. Luke GA, Pathania US, Roulston C et al (2014) DxExNPGP – motives for the motif. Recent Res Devel Virol 9:25–42

20. Halpin C, Cooke SE, Barakate A et al (1999) Self processing 2A-polyproteins – a system for co-ordinate expression of multiple proteins in transgenic plants. Plant J 17(4):453–459
21. Szymczak AL, Workman CJ, Wang Y et al (2004) Correction of multi-gene deficiency *in vivo* using a "self-cleaving" 2A peptide-based retroviral vector. Nat Biotechnol 22 (5):589–594
22. Provost E, Rhee J, Leach SD (2007) Viral 2A peptides allow expression of multiple proteins from a single ORF in transgenic zebrafish embryos. Genesis 45(10):625–629
23. Fang JM, Yi SL, Simmons A et al (2007) An antibody delivery system for regulated expression of therapeutic levels of monoclonal antibodies *in vivo*. Mol Ther 15(6):1153–1159
24. Trichas G, Begbie J, Srinivas S (2008) Use of the viral 2A peptide for bicistronic expression in transgenic mice. BMC Biol 6:40
25. Ha SH, Liang YS, Jung H (2010) Application of two bicistronic systems involving 2A and IRES sequences to the biosynthesis of carotenoids in rice endosperm. Plant Biotechnol J 8 (8):928–938
26. Luke GA, Escuin H, de Felipe P et al (2010) 2A to the fore – research, technology and applications. Biotechnol Genet Eng 26:223–260
27. Kim JH, Lee S-R, Li L-H et al (2011) High cleavage efficiency of a 2A peptide derived from porcine teschovirus-1 in human cell lines, zebrafish and mice. PLoS One 6(4):e18556
28. Samalova M, Fricker M, Moore I (2006) Ratiometric fluorescence-imaging assays of plant membrane traffic using polyproteins. Traffic 7:1701–1723
29. Burén S, Ortega-Villasante C, Őtvős K et al (2012) Use of the foot-and-mouth disease virus 2A peptide co-expression system to study intracellular protein trafficking in *Arabidopsis*. Plos ONE 7(12):e51793. https://doi.org/10.1371/journal.pone.0051973
30. de Felipe P, Luke GA, Brown JD et al (2010) Inhibition of 2A-mediated "cleavage" of certain artificial polyproteins bearing N-terminal signal sequences. Bitotechnol J 5(2):213–223
31. Osborn MJ, Panoskaltsis-Mortari A, McElmurry RT et al (2005) A picornaviral 2A-like sequence-based tricistronic vector allowing for high-level therapeutic gene expression coupled to a dual-reporter system. Mol Ther 12 (3):569–575
32. Gaj T, Gersbach CA, Barbas CF III (2013) ZFN, TALEN, and CRISPR/Cas-based methods for genome engineering. Trends Biotechnol 31(7):397–405
33. Gupta RM, Musunuru K (2014) Expanding the genetic editing tool kit: ZFNs, TALENs, and CRISPR-Cas9. J Clin Invest 124 (10):4154–4161. https://doi.org/10.1172/CJ172992.
34. Ain QU, Chung JY, Kim YH (2015) Current and future delivery systems for engineered nucleases: ZFN, TALEN and RGEN. J Control Release 205:120–127
35. Iliakis G, Wang H, Perrault AR et al (2004) Mechanisms of DNA double strand break repair and chromosome aberration formation. Cytogenet Genome Res 104:14–20. https://doi.org/10.1159/000077461
36. Kim YG, Cha J, Chandrasegaran S (1996) Hybrid restriction enzymes: zinc finger fusions to *FokI* cleavage domain. Proc Natl Acad Sci U S A 93:1156–1160
37. Miller JC, Tan S, Qiao G et al (2011) A TALE nuclease architecture for efficient genome editing. Nat Biotechnol 29:143–148
38. Carroll D (2012) Genome engineering with zinc-finger nucleases. Genetics 188:773–782
39. Fu Y, Foden JA, Khayter C et al (2013) High-frequency off-target mutagenisis induced by CRISPR-Cas nucleases in human cells. Nat Biotechnol 31(9):822–826
40. Hsu PD, Scott DA, Weinstein JA et al (2013) DNA targeting specificity of RNA-guided Cas9 nucleases. Nat Biotechnol 31(9):827–832
41. Ding Q, Lee Y-K, Schaefer EAK et al (2013) A TALEN genome editing system to generate human stem cell-based disease models. Cell Stem Cell 12(2):238–251
42. Xu L, Zhao P, Mariano A et al (2013) Targeted myostatin gene editing in multiple mammalian species directed by a single pair of TALE nucleases. Mol Ther Nucleic Acids 2(7):e112
43. Joglekar AV, Hollis RP, Kuftinec G et al (2013) Integrase-defective lentiviral vectors as a delivery platform for targeted modification of adenosine deaminase locus. Mol Ther 21 (9):1705–1717
44. Mariano A, Xu L, Han R (2014) Highly efficient genome editing *via* 2A-coupled co-expression of two TALEN monomers. BMC Res Notes 7:628
45. Duda K, Lonowski LA, Kofoed-Nielsen M et al (2014) High-efficiency genome editing *via* 2A-coupled co-expression of fluorescent proteins and zinc finger nucleases or CRISPR/Cas9 nickase pairs. Nucleic Acids Res 42(10): e84. https://doi.org/10.1093/nar/gku251
46. Gabriel R, Lombardo A, Miller JC et al (2011) An unbiased genome-wide analysis of zinc-

finger nuclease specificity. Nat Biotechnol 29 (9):816–823

47. Pattanayak V, Ramirez CL, Joung JK et al (2011) Revealing off-target cleavage specificities of zinc-finger nucleases by *in vitro* selection. Nat Methods 8(9):765–770. https://doi.org/10.1038/NMETH.1670

48. Wang W, Ye C, Liu J et al (2014) CCR5 gene disruption *via* lentiviral vectors expressing Cas9 and single guided RNA renders cells resistant to HIV-1 infection. PLoS One 9(12): e115987. https://doi.org/10.1371/journal.pone.0115987

49. Kim H, Kim M-S, Wee G et al (2013) Magnetic separation and antibiotics selection enable enrichment of cells with ZFN/TALEN-induced mutations. PLoS One 8(2):e56476. https://doi.org/10.1371/journal.pone.0056476

50. Hockemeyer D, Wang H, Kiani S et al (2011) Genetic engineering of human pluripotent cells using TALE nucleases. Nat Biotechnol 29 (8):731–734

51. Brandsma I, van Gent DC (2012) Pathway choice in DNA double strand break repair: observations of a balancing act. Genome Integr 3:9. http://www.genomeintegrity.com/content/3/1/9

52. Certo MT, Ryu BY, Annis JE et al (2011) Tracking genome engineering outcome at individual DNA breakpoints. Nat Methods 8 (8):671–676

53. Chu VT, Weber T, Wefers B et al (2015) Increasing the efficiency of homology-directed repair for CRISPR-Cas9-induced precise gene editing in mammalian cells. Nat Biotechnol 33 (5):543–548

54. Luke GA (2012) Translating 2A research into practice. In: Agbo EC (ed) Innovations in biotechnology. InTech, Croatia, pp 161–186

55. Luke GA, Ryan MD (2013) The protein coexpression problem in biotechnology and biomedicine; virus 2A and 2A-like sequences provide a solution. Future Virol 8:983–996

56. Luke GA, Roulston C, Tilsner J et al (2015) Growing uses of 2A in plant biotechnology. In: Ekinci D (ed) Biotechnology. InTech, Croatia, pp 165–193

57. Geier M, Fauland P, Vogl T et al (2015) Compact multi-enzyme pathways in *P. pastoris*. Chem Commun (Camb) 51:1643–1646

58. Ryan MD, Luke GA, Hughes LE et al (2002) The aphtho- and cardiovirus "primary" 2A/2B polyprotein "cleavage". In: Semler BL, Wimmer E (eds) Molecular biology of picornaviruses. ASM Press, Washington, DC, pp 61–70

59. Sharma P, Yan F, Doronina V et al (2012) 2A peptides provide distinct solutions to driving stop-carry on translational recoding. Nucleic Acids Res 40:3143–3151

60. Minskaia E, Nicholson J, Ryan MD (2013) Optimisation of the foot-and-mouth disease virus 2A co-expression system for biomedical applications. BMC Biotechnol 13:67. http://www.biomedcentral.com/1472-6750/13/67

61. Minskaia E, Ryan MD (2013) Protein coexpression using FMDV 2A: effect of "linker" residues. Biomed Res Int 2013:291730. http://dx.doi.org/101155/2013/291730

62. Yang S, Cohen CJ, Peng PD et al (2008) Development of optimal bicistronic lentiviral vectors facilitates high-level TCR gene expression and robust tumor cell recognition. Gene Ther 15(21):1411–1423

63. Tan Y, Liang H, Chen A et al (2010) Coexpression of double or triple copies of the rabies virus glycoprotein gene using a "self-cleaving" 2A peptide-based replication-defective human adenovirus serotype 5 vector. Biologicals 38:586–593

64. Lorens JB, Pearsall DM, Swift SE et al (2004) Stable, stoichiometric delivery of diverse protein functions. J Biochem Biophys Methods 58:101–110

65. Holst J, Vignali KM, Burton AR et al (2006) Rapid analysis of T-cell selection *in vivo* using t cell-receptor retrogenic mice. Nat Methods 3 (3):191–197

66. Ma C, Mitra A (2002) Expressing multiple genes in a single open reading frame with the 2A region of foot-and-mouth disease virus as a linker. Mol Breed 9:191–199

67. Lengler J, Holzmüller H, Salmons B (2005) FMDV-2A sequence and protein arrangement contribute to functionality of CYP2B1-reporter fusion protein. Anal Biochem 343:116–124

68. Rothwell DG, Crossley R, Bridgeman JS et al (2010) Functional expression of secreted proteins from a bicistronic retroviral cassette based on FMDV 2A can be position-dependent. Hum Gene Ther 21(11):1631–1637

69. Appleby SL, Irani Y, Mortimer L et al (2013) Co-expression of a scFv antibody fragment and a reporter protein using lentiviral shuttle plasmid containing a self-processing furin-2A sequence. J Immunol Methods 397:61–65

70. Thomas CL, Maule AJ (2000) Limitations on the use of fused green fluorescent protein to investigate structure-function relationships for the cauliflower mosaic virus movement protein. J Gen Virol 81:1851–1855

71. de Felipe P, Hughes LE, Ryan MD et al (2003) Co-translational, intraribosomal cleavage of polypeptides by the foot-and-mouth disease virus 2A-peptide. J Biol Chem 13:11441–11448
72. Roosien J, Belsham GJ, Ryan MD et al (1990) Synthesis of foot-and-mouth disease virus capsid proteins in insect cells using baculovirus expression vectors. J Gen Virol 71:1703–1711
73. Steiner DF (1998) The proprotein convertases. Curr Opin Chem Biol 2:31–39
74. Fisicaro N, Londrigan SL, Brady JL et al (2011) Versatile co-expression of graft-protective proteins using 2A-linked cassettes. Xenotransplantation 18(2):121–130
75. François IEJA, Hemelrijck WV, Aerts AM (2004) Processing in *Arabidopsis thaliana* of a heterologous polyprotein resulting in differential targeting of the individual plant defensins. Plant Sci 166(1):113–121
76. Sun H, Lang Z, Zhu L et al (2012) Acquiring transgenic tobacco plants with insect resistance and glyphosate by fusion transformation. Plant Cell Rep 31:1877–1887
77. Varshavsky A (1992) The N-end rule. Cell 69:725–735
78. Knoche K, Kephart D (1999) Cloning blunt-end Pfu DNA polymerase-generated PCR fragments into pGEM®-T vector systems. Promega Notes 71:10–13

Chapter 4

Developing Mammalian Cellular Clock Models Using Firefly Luciferase Reporter

Chidambaram Ramanathan and Andrew C. Liu

Abstract

In mammals, many aspects of metabolic, physiological, and behavioral processes are regulated by endogenous circadian clocks. Oscillators of different tissue types share a common molecular mechanism at the cellular and molecular level which underlies the rhythmic expression of genes. Individual cells are the functional units for rhythm generation and cell-based clock models offer experimental tractability for discovery. Cellular clock models can be developed by introducing a noninvasive and readily detectable luciferase bioluminescence reporter as a rhythmic output, in which the promoter of a rhythmically expressed gene is fused with the firefly *luciferase* (*Luc*) gene. The bioluminescence expression in the cells is measured continuously over several days using a highly sensitive and automated recording device. As such, the data are of high temporal resolution and allow precise determination of key circadian parameters including period length, amplitude, damping rate, and phase. Miniaturization of the assays improves throughput for large scale screens. In our lab, we have expertise for constructing circadian reporters and developing reporter cell lines. Here, we describe the procedure for establishing a stable mouse hepatocyte reporter cell line. The procedure described here can be applied to various other cell types.

Key words Circadian clock, Per2, Firefly luciferase, Bioluminescence, Hepatocytes, Mammalian

1 Introduction

In mammals, many aspects of behavior and physiology such as the sleep–wake cycle, body temperature, blood pressure and liver metabolism are regulated by endogenous circadian clocks [1]. The circadian time-keeping system is a hierarchical multioscillator network in which the central clock in the suprachiasmatic nucleus (SCN) synchronizes and coordinates various peripheral oscillators. Individual cells are the functional units for rhythm generation and oscillators of different tissue types share a basic molecular negative feedback mechanism [2]. However, due to strong intercellular coupling in the SCN and confounding factors at the organismal level, delineating gene function in the SCN and in mice has not been straightforward [2, 3]. On the other hand, cell-

Robert Damoiseaux and Samuel Hasson (eds.), *Reporter Gene Assays: Methods and Protocols*, Methods in Molecular Biology, vol. 1755, https://doi.org/10.1007/978-1-4939-7724-6_4,

autonomous clock models lack strong intercellular coupling and reflect what is needed to cycle on an oscillator basis. Thus, cellular clock models are needed to study gene function and mechanisms [2, 3]. Importantly, despite sharing a similar molecular mechanism, oscillators of different tissue types in the organism display cell type-specific characteristics ([4] and references cited therein). Recent studies revealed that ~50% of all genes in the mouse genome oscillate somewhere in the body, often in an organ-specific manner [5]. The circadian oscillators in different tissue types not only regulate different outputs, but are also integrated with and influenced by the local physiologies that are under the control of the clock. Thus, to study cell type-specific clock mechanisms, various cellular clock models representing different physiologies are needed.

Because circadian rhythms are dynamic, longitudinal measurements with high temporal resolution are needed to assess clock function. In recent years, real-time bioluminescence recording using firefly luciferase as a reporter has become a common technique for studying circadian rhythms and probing the underlying mechanisms. Luciferases are naturally occurring oxidative enzymes that catalyze emission of photons from a substrate—luminescence. Luminescence offers advantages over fluorescence. Luciferase luminescence is generated from the cells that catalyze the light-emitting reactions; as such, it does not require exogenous light excitation as in the case of fluorescence and therefore does not cause concerns of phototoxicity or high background levels of light emission. Among the several naturally occurring luciferases, the firefly luciferase has been established as the premier reporter in circadian research (reviewed in [6, 7]). One major reason for the success of the firefly luciferase lies at the superb biochemical and biophysical properties of its substrate, D-luciferin: it is readily dissolvable in water, highly permeable to cells, highly stable for days and even weeks in cells and medium, and is generally not toxic to cells. The successful use of transgenic mice harboring the PER2::LUC reporter attests to the qualifications of the luciferase luminescence reporter [8]. There have been important advancements in bioluminescence recording capabilities, particularly sensitivity and automation. The luciferase reporter, together with the advanced recording devices, has enabled long-term real-time recordings. As such, the bioluminescence data are of high temporal resolution and allow precise determination of key circadian parameters including period length, amplitude, damping rate, and phase, critical for rigorous and systematic phenotypic characterization. It should be noted that these assays can be miniaturized and adapted to high-throughput screening for the discovery of new genes and small molecule modifiers (e.g., [9, 10]).

In our lab, we have constructed various circadian reporters that report different phases of a circadian cycle and can be used for transfection and transduction [3, 11, 12] (*see* **Notes 1–3**). Furthermore, we have developed various reporter cell lines including mouse 3T3 fibroblasts, human U2OS osteosarcoma cells, mouse MMH-D3 hepatocytes and mouse 3T3-L1 adipocytes that express circadian reporters [4, 9]. Here, we describe the procedure for developing stable MMH-D3 mouse hepatocyte reporter cell lines. The procedure described here can be applied to various other cell types and may prove useful in tackling problems in other biological systems. Compared to traditional approaches using mice and tissue explants, these models are more experimentally tractable for phenotypic characterization and rapid discovery of basic clock mechanisms and will prove useful in helping to uncover tissue-ubiquitous, as well as tissue-specific, properties of circadian clocks.

2 Materials

1. Tissue culture facility: various tissue culture vessels including 10-cm, 35-mm, 6-well, 12-well, 24-well and 96-well plates, hemocytometer, microscope with phase contrast and fluorescence capacity, BSL2 certified tissue culture hood, tissue culture incubator (humidified, 37 °C, 5% CO_2), bench top centrifuge (e.g., Eppendorf, Cat. # 5810/5810R).
2. High glucose Dulbecco's modified Eagle medium (DMEM) (HyClone, Cat. # SH30243FS).
3. Fetal Bovine Serum (FBS): need to be tested for certain cells and we have used HyClone serum (Cat. # SH3008803); store at −80 °C for long-term storage.
4. Penicillin-Streptomycin-Glutamine (PSG) (Hyclone, Cat. # SV3008201; 100×): store at −20 °C for long term and 4 °C for daily use.
5. Complete DMEM: add 50 mL FBS and 5 mL 100× PSG in 500 mL of high glucose DMEM; store at 4 °C for daily use.
6. RPMI-1640 Medium (HyClone, Cat. # SH30096FS).
7. Recombinant human insulin (Sigma, Cat. # I9278; ~10 mg/mL, 1000×).
8. Epidermal growth factor (EGF) (Sigma, Cat. # E9644): dissolve 200 μg power in 1 mL DPBS, and store 137 μL aliquots at −20 °C for long-term and 4 °C for short-term use; use at final concentration of 55 ng/mL in culture medium.
9. Insulin like growth factor-II (IGF-II) (Sigma, Cat. # I2526): dissolve 50 μg power in 0.5 mL DPBS in the presence of 0.1% BSA, and store 80 μL aliquots at −20 °C for long-term and 4 °C for short-term use; use at final concentration of 16 ng/mL in culture medium.

10. Complete RPMI medium: add 50 mL FBS and 5 mL 100× PSG in 500 mL of RPMI-1640 medium, supplemented with 0.5 mL insulin (final concentration 10 μg/mL), 137 μL EGF (final 55 ng/mL), and 80 μL IGF-II (final 16 ng/mL); store at 4 °C for daily use.
11. Dimethyl sulfoxide (DMSO) (Sigma, Cat. # D2650): 100%, sterile-filtered, cell culture grade.
12. Dulbecco's Phosphate buffered saline (DPBS) (Hyclone, Cat. # SH30028FS).
13. Trypsin–EDTA mixture (Hyclone, Cat. # SH 30236.01): store at −20 °C for long-term and 4 °C for daily use.
14. Collagen I (MP Biomedicals, Cat. # 160084): make 1 mg/mL stock solution (20×) in 100 mM acetic acid, sterilize through 0.22 μm filter, dilute to 1× working solution (50 μg/mL) in DPBS, and use 1 mL in 35-mm or 4 mL in 10-cm culture dishes for coating at about 5 μg collagen/cm^2.
15. Poly-L-lysine (Sigma, Cat. # P8920): supplied as 0.1% in H_2O (100× and store at 4 °C); dilute in DPBS to make 1× working solution immediately before coating tissue culture plates.
16. Human embryonic kidney (HEK) 293T cells (Invitrogen, Cat. # R700-07).
17. Freezing medium: 90% FBS, 10% DMSO.
18. Packaging vectors of third generation lentiviral system: pMDL-Gag/Pol, pRev, and pVSVG. (Invitrogen, Cat. # K4975-00).
19. 2.5 M $CaCl_2$ stock solution: dissolve 183.7 g of $CaCl_2 \cdot 2H_2O$ (Sigma, Cat. # C7902) in 500 mL of water; sterilize through a 0.22 μm filter; store at −20 °C in 1 mL aliquots.
20. 2× BBS stock solution: dissolve 10.65 g *N*-bis (2-hydroxyethyl)-2-aminoethanesulfonic acid, 16.36 g NaCl, and 0.2 g Na_2HPO_4 in 800 mL of H_2O, adjust to pH 6.95 with 1 N NaOH, and add H_2O to final 1000 mL; sterilize through 0.22 μm filter; store at −20 °C in 10 mL aliquots.
21. Polybrene infection/transfection reagent (Millipore, Cat. # TR-1003-G; 10 mg/mL, 2000×).
22. Blasticidin hydrochloride (Invivogen, Cat. # ant-bl-1): supplied as 10 mg/mL solution in HEPES buffer, pH 7.5; 1–10 μg/mL will be used for selection as empirically determined.
23. Dexamethasone (Sigma, Cat. # D4902): make 2 mM stock solution in 100% ethanol (10,000×); store at −20 °C in 1 mL aliquots.
24. D-Luciferin firefly, potassium salt, (Biosynth, Cat. # L-8220): to make 100 mM stock solution, dissolve 1 g luciferin powder

in 31.4 mL tissue culture grade H_2O, sterilize through 0.22 μm filter, and store 1 mL aliquots.

25. Recording medium: to make 50 mL 1× working solution, add 5 mL DMEM (Invitrogen, Cat. # 12100-046; 10× stock), 1 mL 0.5 M HEPES (H4034; 50×, not buffered stock), 0.21 mL 1 M $NaHCO_3$ (Sigma, Cat. # S5761), 0.26 mL Penicillin-Streptomycin (Invitrogen, Cat. # 15070-063; 5000 units/mL Pen, 5000 μg/mL Strep), 1 mL B-27 supplement (Invitrogen, Cat. # 17504-044; 50×), 0.5 mL 100 mM D-Luciferin, and 42.03 mL tissue culture grade H_2O; adjust to pH 7.4 with 1 N NaOH, sterilize with 0.22 μm filter, and use within a week.
26. Vacuum grease (Fisher Scientific, Cat. # 14-635-5D).
27. 32-channel LumiCycle luminometer (Actimetrics, Inc.).
28. Synergy 2 SL luminescence microplate reader (BioTek, Cat. # 11-120-518).

3 Methods

In this protocol, we describe the use of the lentiviral P(*Per2*)-d*Luc* reporter vector to develop the MMH-D3 mouse hepatocyte clock reporter model. Methods described here can be applied to a great variety of cell types to study the cellular and molecular basis of circadian clocks.

3.1 Construction of Lentiviral Luciferase Reporters

The most commonly used firefly luciferase is *Luc*$^+$ in the pGL3 vector series (Promega), in which the coding region of the native *Luc* was modified for optimized transcription and translation. The destabilized *luciferase* (d*Luc*) is a modified version of *Luc*$^+$ with a PEST sequence fused at its C-terminus to allow for more rapid protein degradation [13]. A circadian reporter usually contains an expression cassette in which the regulatory region of a rhythmically expressed gene is fused with the *Luc* gene. For transient transfection, a circadian promoter can be inserted into the pGL3 vector immediately upstream of the *Luc* gene. For stable integration of the reporter into a host cell's genome, a lentiviral expression vector can be used. Here we describe a method to construct the lentiviral P (*Per2*)-d*Luc* reporter in which the d*Luc* is under the control of the mouse *Per2* promoter.

1. Use PCR to amplify the *Per2* promoter DNA fragment from a mouse *Per2* BAC clone using a forward primer (CTCGAGCGGATTACCGAGGCTGGTCACGTC) and a reverse primer (CTCGAGTCCCTTGCTCGGCCCGTCACTTGG). Clone the 526 bp PCR product into pENTR5′-TOPO vector (Invitrogen) to generate pENTR5′-P(*Per2*).

2. Use PCR to amplify the d*Luc* DNA fragment which contains the firefly *Luc* gene and a C-terminal PEST sequence for rapid protein degradation [12, 13]. Clone the 1776 bp d*Luc* PCR product into pENTR/D-TOPO vector (Invitrogen) to generate pENTR/D-d*Luc*.
3. Mix pENTR5′-P(*Per2*) and pENTR/D-d*Luc* plasmids with the lentiviral destination vector pLV7-Bsd (Bsd, blasticidin resistance gene) [11]. Add LR Clonase II Plus (Invitrogen, Cat. # 12538-013) to the DNA mix and allow the reaction to proceed at room temperature overnight. The recombination reaction will produce the lentiviral pLV7-Bsd-P(*Per2*)-d*Luc* reporter construct (Fig. 1a).
4. Transform the recombination reaction in chemically competent Stbl3 bacterial cells and select single cell clones for plasmid propagation. Isolate the plasmid DNA and confirm the vector by restriction digestion and sequencing.

3.2 Production of Lentiviral Particles of Luciferase Reporters

Luciferase reporters can be introduced into cells via transient transfection or stable transduction (*see* **Note 1**). In this protocol, we focus on stable transduction using lentivirus-mediated gene delivery. We first describe the production of lentiviral particles in 293T cells in 6-well culture plate format. The protocol can be scaled to smaller or larger vessels to produce different amounts of viral particles as needed. It should be noted that aseptic techniques should be used when handling cell cultures and extra caution should be taken when handling cells and medium containing lentiviral particles (*see* **Note 4**).

1. Thaw a fresh tube of 293T cells stored in liquid nitrogen and grow the cells in a tissue culture incubator for 1–2 days.
2. Seed ~5 × 10^6 293T cells on a 10-cm tissue culture plate in 10 mL complete DMEM supplemented with serum and antibiotics, and grow the cells in a tissue culture incubator for 2–3 days. Change medium every other day. Passage the cells when needed and before the cells become confluent. 293T cells can be finicky and require careful handling (*see* **Note 5**).
3. On day 1 (the day prior to transfection), coat a 6-well plate by adding 1 mL of 1× poly-L-lysine to each well and incubate at room temperature for 20 min. Before seeding the cells, aspirate the coating solution and rinse once with DPBS.
4. Split 293T cells with trypsin and seed 0.75 × 10^6 cells per well of the precoated 6-well plate in 2 mL complete DMEM. Ensure that the cells are evenly distributed. Grow the cells overnight.

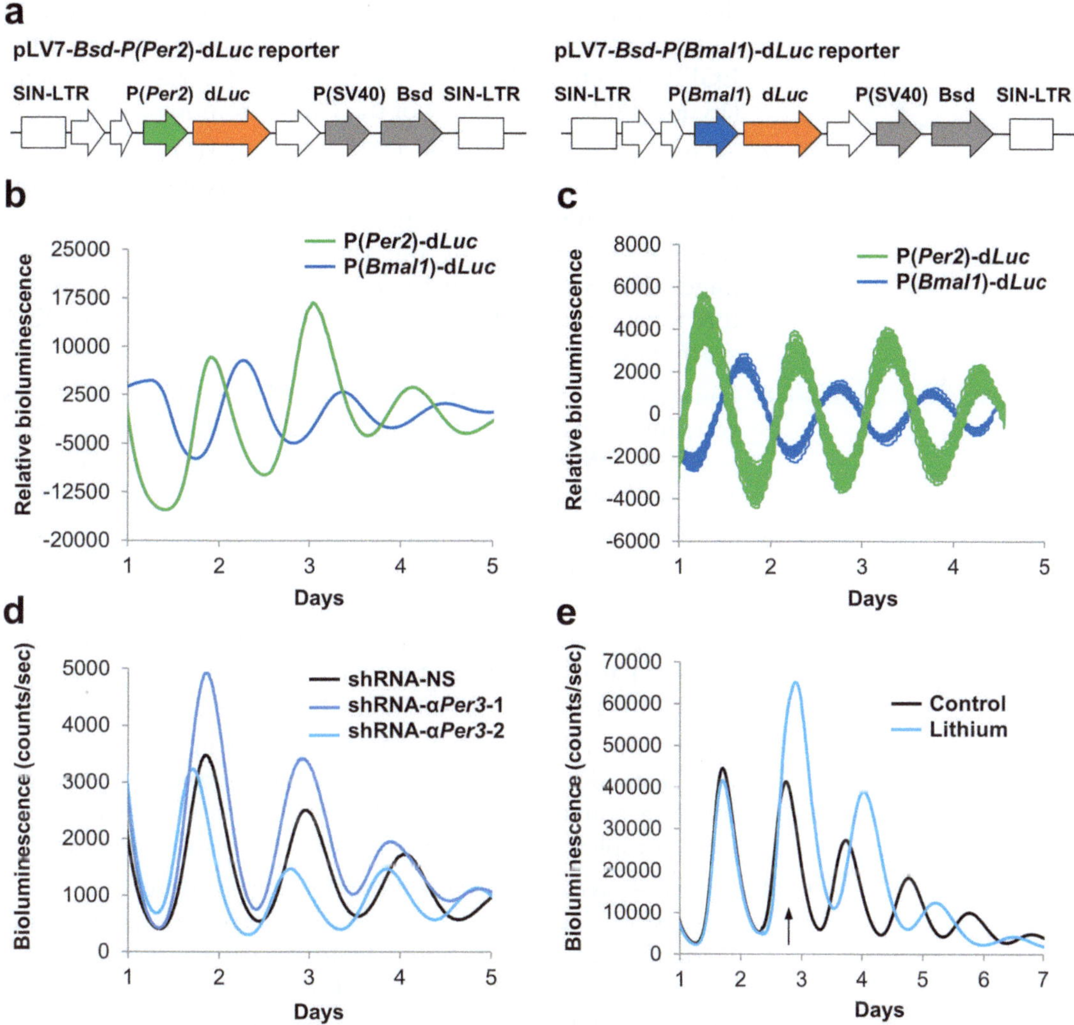

Fig. 1 Establishing MMH-D3 mouse hepatocyte cell clock models using firefly luciferase reporters. (**a**) Diagram of lentiviral P(*Per2*)-d*Luc* and P(*Bmal1*)-d*Luc* reporter constructs. Only the genomic integration region is shown. The d*Luc* expression is under direct control of the mouse *Per2* or *Bmal1* promoters. Blasticidin (Bsd) expression is regulated by the SV40 promoter. Representative bioluminescence rhythms are shown from the LumiCycle recording (**b**) and the Synergy microplate recording (**c**). Note that the two promoters drive antiphasic reporter expression patterns. While knockdown of *Per3* using lentiviral shRNAs shortens (**d**), lithium chloride (20 mM) lengthens (**e**) the period length of MMH-D2 cells expressing the P(*Per2*)-d*Luc* reporter. Lithium was added to the cells at a peak of bioluminescence expression (arrow)

5. On day 2 (the day of transfection), confirm the cell health of the culture under a microscope. Cells should be 80–90% confluent at time of transfection.
6. Prepare the DNA mix for transfection in a 1.5 mL microfuge tube by adding following plasmid DNAs: 2 μg pLV7-P(*Per2*)-d*Luc* as constructed above [11] and 2.5 μg packaging vectors (1.3 μg pMDL-Gag/Pol, 0.5 μg pRev, and 0.7 μg pVSVG) as

described previously [14]. As a control for both transfection and subsequent infection, we usually include pLV156-CMV-EGFP, which harbors enhanced green fluorescent protein (EGFP) under the control of the CMV promoter.

7. To prepare the transfection DNA mix, add 100 μL of 0.25 M $CaCl_2$ into the DNA mix in **step 6** and mix gently but thoroughly by tapping with a finger. Then, add 100 μL of 2× BBS solution and mix gently but thoroughly by tapping with a finger. Incubate the transfection DNA mix at room temperature for 15 min to allow formation of calcium phosphate-DNA precipitates.
8. During incubation, aspirate the DMEM from the 6-well plates containing the 293T cells and replace with 2 mL of fresh complete DMEM medium. Return the plate to the incubator; allow at least 10 min for to equilibrate medium pH before transfection.
9. Take the 6-well plates containing 293T cells from incubator and bring them in the hood. Add the calcium phosphate–DNA precipitates to the cells drop-wise. Swirl the plate gently and observe particle formation under a microscope (*see* **Note 5**). Return the plate to incubator and incubate/grow overnight.
10. On day 3, viral particles start to accumulate in the supernatant (*see* **Note 4**). At least 16 h post-transfection, aspirate the medium containing the transfection mix, replace with 2 mL fresh complete DMEM, and incubate the plate overnight.
11. On day 4, assess the transfection efficiency by observing EGFP expression in transfection control cells (*see* **Note 5**). Collect the medium containing secreted lentiviral particles in a 15 mL centrifuge tube. Centrifuge at >2000 × *g* for 5 min and collect the supernatant containing viral particles. Avoid the pellet which may contain 293T cells.

3.3 Infection of MMH-D3 Hepatocytes with Lentiviral Particles

We will then use the lentiviral particles produced in Subheading 3.2 to infect target cells. The lentiviral system offers unique versatility and high transduction efficiency and can transduce a wide range of cell types and integrate in the host genome in both dividing and nondividing cells. Here, we are using MMH-D3 hepatocytes as a model cell type because of its physiological relevance to liver function, the most well-studied organ in the circadian field.

1. Coat all culture dishes for MMH-D3 cells with 1× collagen I working solution (~5 μg/cm^2). Leave the dish at room temperature under UV inside a hood for 1 h and rinse with DPBS once prior to seeding cells.

2. Grow the cells in complete RPMI medium in collagen-coated culture dishes. Cultures should be observed daily and cells must be passaged at least once before use in any experiment.
3. On day 3 (the day after 293T transfection), passage MMH-D3 cells and seed ~36,000 cells/well onto a collagen-coated 12-well plate. Grow the cells in the incubator overnight.
4. On day 4 (the day of infection), observe the seeded cells. 30–40% confluence is optimal for infection.
5. Add polybrene in a 1:2000 dilution to a final concentration of 5 μg/mL to the collected medium supernatant containing the viral particles (*see* Subheading 3.2). Mix well by pipetting. Caution should be taken from here on when handling lentiviruses (*see* **Note 6**).
6. Aspirate the medium and add 1 mL of the above viral particle mix per well. Incubate the cells in the incubator overnight.
7. On day 5, at least 24 h post-infection, aspirate the medium containing viral particles, wash once with DPBS, and add fresh complete RPMI medium. Incubate the cells in the incubator overnight for recovery and growth.
8. On day 6 and onward, observe cells. When the cells reach confluence, passage to a collagen-coated culture vessel of appropriate size. Grow the cells in the incubator for 1–2 days.
9. When the infected cells reach ~50% confluence, aspirate the medium and add fresh complete RPMI medium containing 10 μg/mL blasticidin to select for stably transduced MMH-D3 cells (*see* **Note** 7). Replenish with fresh complete medium every 2–3 days for continuous selection until control cells are completed killed by the antibiotics.
10. Store some cells from initial passages in freezing medium.
11. Perform bioluminescence recording as described in Subheading 3.5 to examine the circadian properties of the stably transduced cells.

3.4 Selection of Single Cell Clones

Clonal cell lines with genetic homogeneity are often desired and can be obtained by fluorescence-aided cell sorting (FACS)-based single cell sorting. However, this method requires a GFP marker and FACS capabilities, and sorting itself causes stress and reduces cell survival. Here we describe a simple serial dilution method for single cell cloning.

1. Grow the stably transduced MMH-D3 cells on a collagen-coated 10-cm culture dish. Trypsinize the cells before reaching confluence (preferably 50–80%), add 5 mL of complete RPMI medium, pipet up and down several times to further dissociate the cells.

2. Collect the dissociated cells in a 15 mL centrifuge tube and centrifuge at 250 × *g* for 5 min.
3. Aspirate the medium, add 10 mL complete RPMI medium, and resuspend the pelleted cells by gentle and thorough pipetting up and down >10 times to ensure complete dissociation and single cell suspension.
4. Count cell number using a hemocytometer and make serial dilutions to final 1 cell/150 μL of medium. Distribute 150 μL of the diluted cell suspension onto 96-well plates using a multichannel pipette.
5. Incubate the 96-well plates for 2–3 weeks and change to fresh medium once a week. Observe every well under a phase contrast microscope and look for single cell colony formation. Discard those wells that show apparent formation of multiple colonies.
6. When cells in individual wells reach confluence, trypsinize the single cell colony and passage to 24- or 12-well plates. In our experience, six 96-well plates should give a few dozen single cell clones for further characterization.
7. Examine the morphology and growth rates of the individual clonal cell lines, both prehepatocytes and mature hepatocytes upon differentiation (*see* Subheading 3.5) to make sure that they comparable to parental MMH-D3 cells.
8. Store some cells from initial passages in freezing medium.
9. Perform bioluminescence recording as described in Subheading 3.5 to examine the circadian properties of clonal cell lines. Brighter cells can be obtained from screening of the clonal lines (*see* **Note 8**).

3.5 Bioluminescence Recording of Luciferase Reporter Cells

The circadian rhythms of the stably transduced reporter cells can be examined through bioluminescence recording either using LumiCycle on 35-mm dishes (described here) or using Synergy 2 SL luminescence microplate reader on 96-well plates (*see* **Note 9**). To examine physiologically relevant circadian functions, the MMH-D3 prehepatocytes must first be differentiated into mature hepatocytes prior to bioluminescence recording.

1. Grow the reporter cells in a 10-cm collagen-coated plate until ~90% confluence and passage to coated 35-mm culture dishes. We usually test ≥3 dishes for each line under each condition. (For Synergy multiwell recording, collagen-coated 96-well plates will be used.)
2. After the MMH-D3 prehepatocytes become confluent, change to fresh medium and continue to incubate for 2 days.

3. Aspirate the medium and change to complete RPMI medium supplemented with 2% DMSO (differentiation medium) to initiate cell differentiation.
4. Change fresh differentiation medium every 2 days for 8–10 days to allow cells to fully differentiate into mature hepatocytes. Observe cell morphology changes over the course of differentiation [4, 15].
5. After differentiation, aspirate the medium, rinse once with DPBS, and replace with 2 mL serum-free RPMI medium containing 200 nM dexamethasone (synchronization medium). Incubate in the tissue culture incubator for 2 h to synchronize the cells.
6. Two hours later, aspirate the synchronization medium, rinse once with DPBS, and replace with 2 mL of HEPES-buffered recording medium containing 1× B-27 and 1 mM luciferin. Optimization of recording medium may be necessary (*see* **Note 10**).
7. Following recording medium change, cover the 35-mm culture dishes with 40-mm sterile coverslips and seal in place with vacuum grease to prevent evaporation (Similarly, 96-well plates can be sealed in place with the lid.).
8. Load the sealed 35-mm dishes in the LumiCycle luminometer, which is kept inside an unhumidified incubator at 36 °C without CO_2 or H_2O (96-well plate is loaded in the Synergy microplate reader.).
9. Start real-time bioluminescence recording. We usually record bioluminescence for 1 week, followed by a change to fresh recording medium and continuous recording for a second week. For the LumiCycle assay, the cells are sampled once in 10 min with a 70 s integration time. (For the Synergy assay, 30 min interval and 15 s integration.)
10. Once a cell line is established, it can be used in various mechanistic studies (*see* **Note 11**).

3.6 Data Analysis

The high temporal resolution luminescence data obtained above are critical for determining circadian parameters including period length, amplitude, damping rate and phase. We use the LumiCycle Analysis program (version 2.31, Actimetrics) to analyze bioluminescence data from the LumiCycle assay (described here), and the MultiCycle Analysis program (version 2.39, Actimetrics) to analyze the Synergy data [4, 11].

1. Open data files using the LumiCycle Analysis program. Raw data are baseline fitted first and baseline-subtracted data are fitted to a sine wave (damped), from which circadian parameters including period length, goodness-of-fit and damping

constant are determined. The polynomial number needs to be adjusted during fitting. Due to high transient luminescence upon medium change, we usually exclude the first cycle of data from analysis. For samples that show persistent rhythms, goodness-of-fit of >80% is usually achieved.

2. For amplitude analysis, set polynomial number = 1 and fit the baseline-subtracted data from days 3 to 5 of a 7-day recording to a sine wave, from which the amplitude was determined.
3. For each analysis, export raw data, baseline-subtracted data, and fit parameters.
4. Open the exported data using Excel and plot raw data (bioluminescence, counts/s) against time (days). For direct comparison of phase and amplitude under different perturbation conditions, we usually plot baseline-subtracted data (relative bioluminescence) against time (days) (Fig. 1b, c).
5. Perform statistical analysis of fit parameters to obtain period length, goodness-of-fit, damping rate (1/damping constant), amplitude, and relevant statistical significance.

4 Notes

1. Gene delivery method considerations. Vector-based bioluminescence reporters can be introduced into various cell types via transient transfection or stable transduction. Compared to traditional methods such as transient transfection and germline transmission, the lentiviral vector system is preferred because of the stable integration into the cell's genome and greater transduction efficiency and versatility in dividing and nondividing cells of a great variety of cell types. As such, it is largely not limited to cell types. Although cumbersome and less reproducible, transiently transfected cells can also be grown in the presence of antibiotics to generate stable cell lines.
2. Vector considerations. For transient transfection, the pGL3 vector series harboring *Luc*$^{+}$ is preferred. The pGL4 vector series harboring *Luc2* can also be considered. In our experience, the pLV7 lentiviral vector series usually gives satisfactory results. Depending on needs, other vectors can also be considered. For example, we have used the pLV156-SV40(P/T)-P (*Per2*)-d*Luc*-IRES-EGFP vector in our studies [3], in which the P(*Per2*)-d*Luc* is the circadian reporter and EGFP translation is mediated by an internal ribosome entry site (IRES) downstream of d*Luc*, allowing for visual observation and FACS sorting. The SV40 constitutive promoter/terminator (P/T) serves as an insulator or decoy to shield the reporter from integration site effects.

3. Circadian phase-specific reporters. The core clock mechanism is based on an E/E′-box-mediated transcriptional negative feedback mechanism. The core loop regulates and integrates with two other circadian elements, the D-box and the RRE. The above three circadian elements give rise to three distinct circadian phases or peak expression levels at different times of day, represented by *Per2*, *Per3*, and *Bmal1* promoters, respectively. Combinatorial regulation can generate novel intermediate phases. For example, *Cry1* transcription is mediated by all three circadian elements, giving rise to the distinct *Cry1* phase [16]. Accordingly, at least four distinct phases can be recapitulated in cell-based reporter assays [11]. Thus, based on the regulatory mechanism under study, a phase-specific reporter can be considered.
4. The most commonly used lentiviral vectors include several safety features engineered to enhance biosafety, namely self-inactivation of the lentivirus after transduction of target cells. The lentiviral particles produced in this system are replication incompetent. However, all handling, storage, and disposal of lentiviral particles must be in accordance with institute rules and regulations. Proper training on relevant policies and procedures is required.
5. High transfection efficiency of HEK 293T cells is critical for a lentiviral preparation. Transfection efficiency relies mainly on two factors: cell health and transfection reagents. (a) 293T cells are commonly used for production of lentiviral particles. 293T cells can be finicky and require careful handling. It is highly recommended that low passage number and rapidly growing cells be used for viral production. Cell growth rate and morphology must be carefully monitored. For consistency, we always test different batches of serum and use the same batch for growing 293T cells. We never allow 293T cells to grow over confluent. (b) We usually use the calcium phosphate transfection method. We always test new 2× BBS and 2.5 M $CaCl_2$ stock solutions and optimize working concentrations of $CaCl_2$ to 0.25 M. 293T cells are readily transfectable with a number of transfection reagents. Fine particle formation of calcium phosphate-DNA precipitate is critical for efficient transfection. Transfection efficiency of 90–100% as estimated by EGFP expression is a reliable predictor of a good viral preparation. In addition to calcium phosphate, we also achieved excellent transfection efficiency with Fugene 6 (Roche or Promega), polyethylenimine or PEI (Polysciences), or Lipofectamine-2000 (Invitrogen). These lipid-based transfection (lipofection) reagents are more expensive, but give better transfection consistency than calcium phosphate.

6. Polybrene is used to enhance infection efficiency, but is not absolutely required. As it may be toxic to some cells, prior testing is recommended.
7. The concentration of the antibiotics used for selection needs to be empirically determined for each cell line before the experiment is undertaken. Plot a kill-curve for each cell line by measuring cell growth/death under increasing concentrations of the antibiotics. For the lentiviral vector system that results in genomic integration of the antibiotic resistance gene, the optimal concentration is one that kills the control cells within a week.
8. For high-throughput screening assays using less sensitive recording devices, cells with brighter or higher bioluminescence expression levels are often needed. For this, larger culture vessels such as 15-cm dishes can be used for transfection to produce large quantities of viral particles and obtain high titer viral particles through ultracentrifugation [14]. These viral particles are then used to infect target cells to increase the number of integration sites. Selection of brighter single cell clones from the population of the reporter cells is highly recommended.
9. Recording devices. LumiCycle (Actimetrics) and Kronos (Atto Co.) are the two most commonly used automated luminometer device for real-time recording. Both devices employ photomultiplier tubes (PMTs) as light detectors, which provide extremely high sensitivity and low noise and are particularly suitable for data acquisition of extremely dim luciferase-based bioluminescence. For high-throughput screening experiments, highly sensitive recording devices that accommodate multiwell plates can be tested as described previously [11]. For this, integration/exposure time and interval between time points must be determined empirically for a given reporter and recording device. Further, to obtain spatial information, the cells can be used in single cell imaging using a specially designed microscope with a highly sensitive, low-noise CCD camera [6].
10. Recording medium may need to be optimized, depending cell lines and cell types, to improve cell health and persistence of cellular rhythms. Major considerations include the need for and amounts of B-27, serum and forskolin, and the concentration of $NaHCO_3$ and luciferin [11]. 1 mM luciferin is usually sufficient, but a final concentration (0.1–1 mM) may be determined for each cell/reporter type.
11. Use of cellular clock models. Real-time bioluminescence recording has been conducted in cells and tissue explants derived from mice. Unlike tissue or animal models, cell-based

models are amenable to genetic and pharmacologic perturbations, and when necessary, also use in high-throughput formats. The relative ease of manipulation and scalability of cell culture allows for drastic reduction of the number of experimental animals, which eliminates constrains related to animal use and improves statistical power. Perturbation of gene function can be achieved by overexpression or RNAi-mediated knockdown. Selective small molecules can be used to interfere with protein function. Owing to the inherent high temporal resolution and experimental tractability, the cell-autonomous clock models have greatly facilitated mechanistic studies (e.g., [3, 16–18]. Adaptation to high-throughput recording systems has allowed genome-wide screening of siRNA and shRNA libraries for identification of novel clock factors (Fig. 1d) (e.g., [9, 19], and screening for diverse small molecules that impact clock function (Fig. 1e) (e.g., [10, 20]). By establishing and characterizing more cell type specific clock models, future studies are expected to uncover tissue specific clock properties underlying local circadian biology.

References

1. Hastings MH, Reddy AB, Maywood ES (2003) A clockwork web: circadian timing in brain and periphery, in health and disease. Nat Rev Neurosci 4:649–661
2. Mohawk JA, Green CB, Takahashi JS (2012) Central and peripheral circadian clocks in mammals. Annu Rev Neurosci 35:445–462
3. Liu AC, Welsh DK, Ko CH et al (2007) Intercellular coupling confers robustness against mutations in the SCN circadian clock network. Cell 129:605–616
4. Ramanathan C, Xu H, Khan SK et al (2014) Cell type-specific functions of period genes revealed by novel adipocyte and hepatocyte circadian clock models. PLoS Genet 10: e1004244
5. Zhang R, Lahens NF, Ballance HI et al (2014) A circadian gene expression atlas in mammals: implications for biology and medicine. Proc Natl Acad Sci U S A 111:16219–16224
6. Welsh DK, Imaizumi T, Kay SA (2005) Real-time reporting of circadian-regulated gene expression by luciferase imaging in plants and mammalian cells. Methods Enzymol 393:269–288
7. Yamazaki S, Takahashi JS (2005) Real-time luminescence reporting of circadian gene expression in mammals. Elsevier, Amsterdam
8. Yoo SH, Yamazaki S, Lowrey PL et al (2004) PERIOD2::LUCIFERASE real-time reporting of circadian dynamics reveals persistent circadian oscillations in mouse peripheral tissues. Proc Natl Acad Sci U S A 101:5339–5346
9. Zhang EE, Liu AC, Hirota T et al (2009) A genome-wide RNAi screen for modifiers of the circadian clock in human cells. Cell 139:199–210
10. Hirota T, Lee JW, St John PC et al (2012) Identification of small molecule activators of cryptochrome. Science 337:1094–1097
11. Ramanathan C, Khan SK, Kathale ND et al (2012) Monitoring cell-autonomous circadian clock rhythms of gene expression using luciferase bioluminescence reporters. J Vis Exp 67: e4234
12. Liu AC, Tran HG, Zhang EE et al (2008) Redundant function of REV-ERBalpha and beta and non-essential role for Bmal1 cycling in transcriptional regulation of intracellular circadian rhythms. PLoS Genet 4:e1000023
13. Ueda HR, Chen W, Adachi A et al (2002) A transcription factor response element for gene expression during circadian night. Nature 418:534–539
14. Tiscornia G, Singer O, Verma IM (2006) Production and purification of lentiviral vectors. Nat Protoc 1:241–245
15. Amicone L, Spagnoli FM, Späth G et al (1997) Transgenic expression in the liver of truncated

Met blocks apoptosis and permits immortalization of hepatocytes. EMBO J 16:495–503

16. Ukai-Tadenuma M, Yamada RG, Xu H et al (2011) Delay in feedback repression by cryptochrome 1 is required for circadian clock function. Cell 144:268–281
17. Sato TK, Yamada RG, Ukai H et al (2006) Feedback repression is required for mammalian circadian clock function. Nat Genet 38:312–319
18. Baggs JE, Price TS, DiTacchio L et al (2009) Network features of the mammalian circadian clock. PLoS Biol 7:e52
19. Maier B, Wendt S, Vanselow JT et al (2009) A large-scale functional RNAi screen reveals a role for CK2 in the mammalian circadian clock. Genes Dev 23:708–718
20. Chen Z, Yoo S-H, Park Y-S et al (2012) Identification of diverse modulators of central and peripheral circadian clocks by high-throughput chemical screening. Proc Natl Acad Sci U S A 109:101–106

Chapter 5

High-Throughput Screening Method to Identify Alternative Splicing Regulators

Cheryl Stork and Sika Zheng

Abstract

Misregulation of alternative pre-mRNA splicing contributes to various diseases. Understanding how alternative splicing is regulated paves the way to modulating or correcting molecular pathogenesis of the diseases. Alternative splicing is typically regulated by ***trans*** RNA binding proteins and their upstream modulators. Identification of these splicing regulators has been difficult and traditionally done piecemeal. High-throughput screening strategies to find multiple regulators of exon splicing have great potential to accelerate the discovery process, but typically confront low sensitivity and specificity of splicing assays. Here we describe a high-throughput screening method using dual-fluorescence minigene reporters to allow for sensitive detection of exon splicing changes. To enhance specificity we introduce two complementary dual-fluorescence minigenes that each express both GFP and RFP in response to exon inclusion and exclusion but oppositely. The method significantly eliminates false positives and allows for sensitive identification of true regulators of splicing. The method described here is designed to screen cDNA libraries, but can be applied to isolate splicing regulators from shRNA libraries or chemical libraries.

Key words Dual-fluorescence minigene reporters, High-throughput screen, Alternative splicing, GFP, RFP

1 Introduction

Most mammalian genes produce multiple mRNA isoforms due to pre-mRNA alternative splicing. This process is controlled by a multitude of regulatory factors. However identifying these factors is both laborious and previously would have needed to be done individually. Here, we describe a high throughput screening (HTS) method that allows for identification of multiple positive and negative regulators of an exon of interest. In this method, two minigene reporters are constructed: one that produces GFP when an exon is included and RFP from the mRNA lacking the exon and a second one with RFP and GFP to represent exon inclusion and exclusion respectively. Combining the screening results from these two

Robert Damoiseaux and Samuel Hasson (eds.), *Reporter Gene Assays: Methods and Protocols*, Methods in Molecular Biology, vol. 1755, https://doi.org/10.1007/978-1-4939-7724-6_5, © Springer Science+Business Media, LLC, part of Springer Nature 2018

reporters significantly eliminates false positives and enriches for identification of true splicing regulators.

Pre-mRNA splicing patterns are highly regulated to produce functionally distinct gene products during development or in response to extracellular stimuli [1, 2]. Interactions between *cis*-regulatory elements in the pre-mRNA and *trans*-acting protein factors can affect spliceosome assembly and splice site selection. A single alternative exon can contain multiple cis-elements in itself or in its surrounding introns, and thus can be regulated by multiple trans-factors. Current genome-wide methods allow for identification of many targets of individual splicing factors, however identifying individual positive and negative regulators of splicing has been challenging [3–5]. Previous strategies of expression cloning and various cell-based screening techniques were able to identify few new factors that target exons however were overall insensitive for identifying a larger sets of regulators of splicing [6–9].

Single-output splicing reporters usually measure the overall expression, but not the splicing, of the reporter transcript directly. Additionally, the alternative isoform must be produced at a much lower basal level for any changes in splicing to be measurable in the reporter output. The use of single-output splicing reporters usually screens for activators and repressors of splicing, but not both. The use of a pair of single-output reporters that encode two different fluorescent proteins can overcome some of these limitations [10]. However, possible integration of minigenes containing two different reporter genes into different genomic loci with different copy numbers can impose difficulty in analyzing results.

A dual-output reporter, in which both splicing isoforms are assayed, allows for screening of changes in isoform ratio, and can detect both increases and decreases in exon inclusion with high sensitivity. This screen allows for identification of isoforms whose basal splicing is at intermediate levels and can also reduce the amount of false positives that alter overall reporter expression independently of splicing. Previously, a dual-fluorescence splicing reporter was used in a forward genetic screen in *Caenorhabditis elegans* [11]. The reporter required insertion of the open reading frame (ORF) of green fluorescent protein (GFP) and red fluorescent protein (RFP) into each of the mutually exclusive exons. However, this technique may not work in mammalian cassette exons because differences in translation and stability of the two protein products can produce false positives [12].

The development of cataloged libraries and high-throughput robotic systems has enabled screening at greater depth and the identification of multiple regulators from a single screen [13, 14]. Here we present a broadly applicable cell-based high-throughput screening (HTS) method to simultaneously identify multiple activators and repressors of an alternative exon. Using two dual-fluorescence minigene reporters we minimized systematic

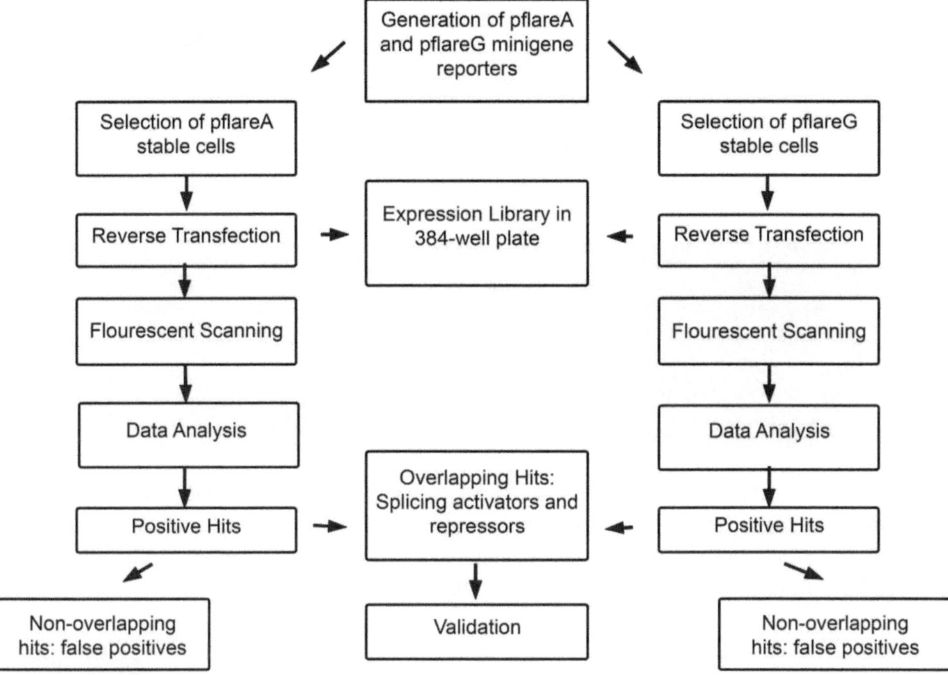

Fig. 1 The framework of the cell-based high-throughput screens to identify splicing factors

variations associated with fluorescent screening to enable more sensitive and accurate detection of moderate splicing changes. *See* Fig. 1 for a general overview of the overall strategy. The first reporter, pflareA, is constructed by splitting the GFP ORF between two constitutive exons by the alternative cassette exon of interest with its flanking introns (Fig. 2). The GFP ORF is initiated for translation when the alternative exon is skipped and the RFP will be silenced. When the alternative exon is included, the GFP ORF loses its start codon and the downstream RFP is then translated.

Some alternative exons may have in-frame start codon for the GFP reading frame. In that case, ATG start codons within the alternative exon need to be mutated. The pflareG minigene reporter is constructed by having the start codon of GFP in the alternative exon such that when the alternative exon is included GFP is translated (Fig. 3) [15]. This can be easily done if the alternative exon has an in-frame start codon. Otherwise an ATG start codon for GFP within the alternative exon can be engineered by site-directed mutagenesis or insertion. When the alternative exon is skipped, GFP does not have an initiation codon and the downstream RFP ORF is used. Overall the method described here greatly reduces false positives while maintaining high sensitivity in detecting regulators. This system allows for genome-wide screening for factors regulating splicing of an exon of interest.

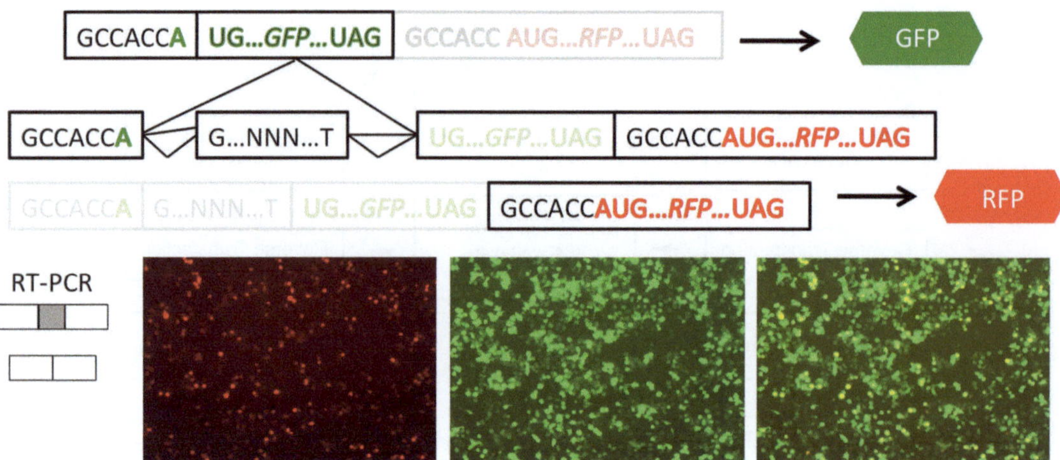

Fig. 2 pflareA minigene reporter: The pflareA minigene contains GFP and RFP open reading frames and an alternative exon of interest. The GFP start codon is split between two consecutive exons that flank the exon of interest. When the exon is skipped the GFP ORF is initiated and GFP is expressed. When the exon is included, the GFP loses its start codon and RFP translation is initiated and expressed

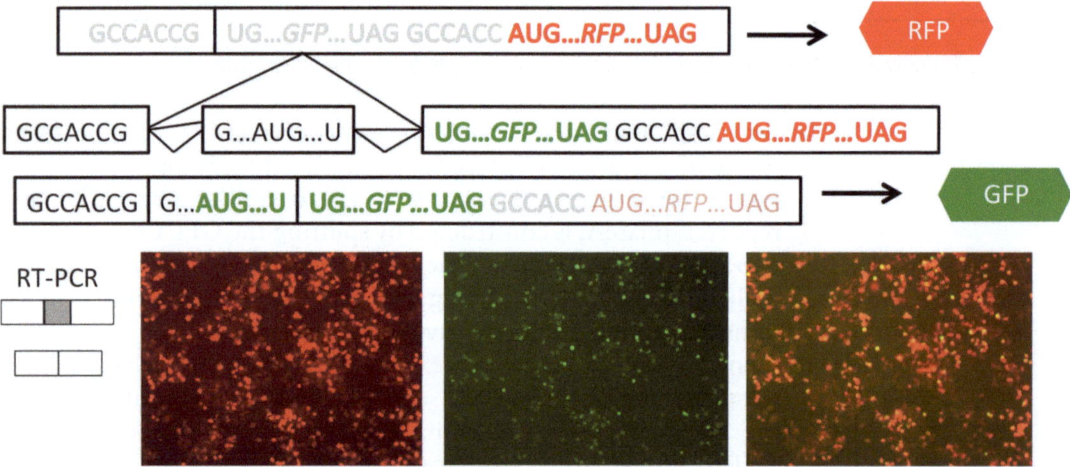

Fig. 3 pflareG minigene reporter: The start codon of GFP ORF is within the alternative exon and GFP measures exon inclusion. When the exon is skipped GFP loses its start codon and initiation of RFP occurs. RFP expression represents exon exclusion

2 Materials

2.1 Construction of Dual-Flourescence Minigene Reporters

1. pflare G vector.
2. pflare A vector.
3. Quickchange site-directed mutagenesis kit (Stratagene).
4. Dra III restriction enzyme.
5. Bam HI.

6. Cultured cell line.
7. G418.

2.2 Library and Array Construction

1. Mammalian Gene Collection (MGC) based assay ready complementary DNA (cDNA) (Open Biosystems).
2. pCMVsport6.0 vector (Life Technologies).
3. pCMVsport6.1 vectors (Life Technologies).
4. cDNA arrayed from MGC IMAGE IRAK sourced plates.
5. 384-well plates.
6. Genetix Qbot (Molecular Devices).
7. Plasmid Mini-Prep Consumables.
8. Biomek FX robot (Beckman Coulter).

2.3 High-Throughput Screens

1. Opti-MEM.
2. Multidrop 384 (Thermo Labsystems).
3. Lipofectamine 2000 (Life Technologies).

2.4 Validation

1. Lipofectamine 2000 (Life Technologies).
2. Silencer Select siRNA (Life Technologies).
3. Lipofectamine RNAiMax (Life Technologies).
4. TRIzol (Life Technologies).
5. Superscript III (Life Technologies).
6. 8% Urea-PAGE gels.

3 Methods

Carry out all procedures at room temperature unless otherwise specified. For commercial kits and reagents, follow manufacturer's protocol.

3.1 Construction of Dual-Fluorescence Minigene Reporters

1. To construct the minigene reporters, the alternative exon together with its flanking intronic sequences will need to be inserted into the EcoRI and BamHI sites of either the pflareA or pflareG vectors. The length of inserted intronic sequence can be optimized. Conserved intronic sequence is usually indicative of regulation and is recommended to be included [16]. In many cases, different length of intronic sequences is chosen to generate multiple versions of minigenes. The one with a mid-range inclusion ratio is often used for screening.
2. When designing the pflareA reporter, any ATG start codons present within the alternative exon of interest will need to be

mutated using the QuickChange site-directed mutagenesis kit following the manufacturer's protocol (*see* **Note 1**).

3. When designing the pflareG reporter, an ATG start codon within the alternative exon in frame with the GFP ORF is required. If missing, a start codon needs to be created by site-directed mutagenesis or insertion (*see* **Note 2**).
4. After ATG generation/removal, linearize the resulting pflareA and pflareG minigene reporters with DraIII and transfect into the target cell line of choice using Lipofectamine 2000 following manufacturer's protocol to start generating stable cell lines.

3.2 Selection of Stable Cell Clones

1. Begin selection of stably transfected cell lines using the positive selection marker G418. Incubate transfected cells in G418-containing media for 2 weeks. Split the cells when needed. Maintain cells in media without phenol red for fluorescence visualization and imaging.
2. FACS sort single stable cell clones that express GFP and RFP into 96-well plates. Grow and expand the cell clones to 12-well plates. Some Clones may lose GFP and RFP expression during the selection process and can be discarded.
3. The stable cell clones need to be tested for their responsiveness to splicing changes before the high-throughput screen. An appropriate clone should not change GFP/RFP ratio when transfected with an empty control vector, and should do so when transfected with a positive control gene (if available). A positive control can be a known activator or repressor. An RT-PCR assay to measure the splicing changes at the RNA level is highly recommended. This step will narrow down the number of appropriate cell clones.
4. Test the "transfectability" of each selected stable cell clone using a GFP expression plasmid. A highly transfectable cell clone would show very high GFP signals without affecting the RFP signals.
5. Using the GFP expression plasmid, use different transfection reagents, transfection reagent to DNA ratios, and cell densities of selected clones to optimize the final transfection conditions for your screen.
6. Select the optimal stable cell clones from **step 4** and grow in larger scale.

3.3 Library and Array Construction for High-Throughput Screens

1. Prepare an "assay ready" cDNA library by duplicating the MGC collection (Open Biosystems) in the pCMVsport6.0 and pCMVsport6.1 vectors prearrayed in 96-well plates (Life Technologies).

2. Array cDNAs from MGC IMAGE IRAK source plates into 384-well plates using a Genetix Qbot (Molecular Devices). A library of about 16,000 clones would occupy 45 384-well plates. Prepare plasmid DNA using plasmid prep consumables (Macherey-Nagel) on a Biomek FX robot (Beckman Coulter), normalized, and spotted into assay plates for screening. We typically prespot 40 ng MGC cDNA per well (except A-H23 and A-H24) in each 384-well plate.
3. We typically spot 40 ng negative control plasmids to wells of C23, C24, D23, and D24. We also spot 40 ng plasmids of positive regulators to wells of E23, E24, F23, and F24, as well as negative regulators to wells of G23, G24, H23, and H24 [16].

3.4 High-Throughput Screens

1. Premix lipofectamine with opti-MEM. Dispense 10 μL opti-MEM containing 0.12 μL Lipofectamine 2,000 in each well to mix with the plasmid DNA prearrayed in the plates.
2. Incubate the plates for 25 min.
3. Add 8,000 reporter cells to every well except A23, A24, B23, and B24 wells. Dispense cell media into these four wells to derive a scale factor for normalization (*see* **Note 3**).
4. Include background plates containing media alone for inter-plate background subtraction.

3.5 Data Acquisition and Background

1. To examine the effect of pixel size on data quality you will need to measure the consistency of signal intensities acquired at different pixel sizes. We often find a pixel size of 200 μm provides both quality data and fast screen. At a pixel size of 200 μm, the 44-plate MGC library can be scanned within 100 min for both GFP and RFP channels (*see* **Note 4**).
2. Add uniform media to a 384-well plate and measure well intensities at different positions for the scanner to test the uniformity of measured fluorescence signals across scanning areas. Exclude areas that show high deviation from the means.
3. Forty-eight hours after transfection, arrange the plates on the Typhoon scanner and separately obtain GFP and RFP signals of every well. Keep the parameters for the scanning consistent across all plates.
4. To correct interplate variation, first identify the GFP and RFP signal intensity of each of the four media-alone wells (A23, A24, B23, and B24) in all plates including the cell plates and the background plate. For example XFP_{ijk} is the raw GFP or RFP intensities of a well at row "*i*" column "*j*" in plate *k*. Calculate the scale factor S_k for plate *k* based on the A23, A24, B23, and B24 wells by the following equation:

$S_k = \frac{1}{4}\sum\frac{XYP_{pq}}{XYP_{pqk}}$ (p = A, B and q = $23, 24$), where XYP_{pq} is the corresponding average values across all the plates.

5. The normalized GFP and RFP expression values (XFP_x) of the stable cells after transfection with $cDNA_x$ at row "*i*" column "*j*" in plate *c* are calculated with the following equation: $XFP_x = XFP_{ijc}S_c - \frac{1}{n_b}\sum_{k=1}^{n_b} XYP_{ijb}S_b$, where S_c and S_b are the scale factors for the cell plates and the background plates respectively, and n_b is the total number of background plates used.
6. To calculate the splicing ratio of the reporter upon expression of $cDNA_x$ use the $\frac{GFP_x}{RFP_x}$ for the pflareG minigene and $\frac{RFP_x}{GFP_x}$ for the pflareA minigene.
7. To calculate the basal splicing level of the reporter at a given time use the four wells that were transfected with empty vector in every plate by calculating the mean $\frac{GFP_{ctrl}}{RFP_{ctrl}}$ in the pflareG-exon cells and the $\frac{RFP_{ctrl}}{GFP_{ctrl}}$ in the pflareA-exon cells.
8. To calculate the action of $cDNA_x$ on the splicing of the pflareG reporter, the change in the splicing ratio (M_x) is calculated as:

$$M_x = \log_2\left[\frac{GFP_x}{RFP_x}\right] - \log_2\left[\frac{GFP_{ctrl}}{RFP_{ctrl}}\right]$$

where $\frac{GFP_{ctrl}}{RFP_{ctrl}}$ was derived from the same plate as $cDNA_x$.

9. To calculate the action of $cDNA_x$ on the splicing of the pflareA reporter, the change in the splicing ratio (M_x) is calculated as:

$$M_x = \log_2\left[\frac{RFP_x}{GFP_x}\right] - \log_2\left[\frac{RFP_{ctrl}}{GFP_{ctrl}}\right]$$

where $\frac{GFP_{ctrl}}{RFP_{ctrl}}$ was derived from the same plate as $cDNA_x$.

10. A value of $M_x > 0$ will indicate a possible increase in splicing by $cDNA_x$. A value of $M_x < 0$ indicates a possible decrease in splicing.
11. To calculate the average fluorescence intensity of cells after transfection, or the "*A*" value, use the following equation:

$$A_x = \frac{1}{2}(\log_2(GFP_x) + \log_2(RFP_x))$$

12. Apply local FDR control to determine the cutoff of the *M* values in calling a positive hit using the R package locfdr.

3.6 Validation

1. Overlap the hits from the two screens with the pflareG and pflareA minigenes and filter out hits identified from only one of the screens.
2. To validate the overlapped hits, perform cDNA plasmid overexpression or RNAi mediated knockdown in a naïve cell line. For example, transfect cells with 1 μM siRNA using Lipofectamine RNAiMax or cDNA plasmids with Lipofectamine 2000 following manufacturer's protocol.
3. Incubate for 48 h to allow cells to express cDNA plasmids or 48–72 h for the RNAi-mediated knockdown.
4. Extract total RNA using TRIzol following manufacturer's protocol.
5. Perform reverse transcription using Superscript III.
6. Perform quantitative or semiquantitative PCR to measure the included and skipped isoforms.
7. Run PCR products on 8% Urea-PAGE gels.
8. Image gels on a Typhoon Imager.
9. Quantify the band intensities using ImageQuant TL and calculate the inclusion ratio of the alternative exon under control, overexpression and knockdown conditions.
10. Use the cutoff of at least 5% change in the inclusion ratio (ΔPSI) from negative control with a *P*-value of <0.05 to confirm true splicing regulators.

4 Notes

1. Mutating the ATG codons present in the alternative exon allows for pflareA-exon of interest to express RFP when the exon is included. Skipping the exon in pflareA generates an AUG codon for GFP.
2. Start codons in the exon of interest that are in frame with the GFP ORF of pflareG are required for expression of GFP. When the exon is excluded the GFP ORF will not be translated and RFP will be expressed.
3. The amount of stable cells and transfection reagent used for the screen will need to be optimized prior to the experiment.
4. Image acquisition must be rapid enough to minimize changes in GFP/RFP during the scan of the whole library. The speed of scanning is inversely correlated with the pixel size of the acquired image. An optimal scan should take the shortest time but still produce an image resolution of sufficient data quality.

References

1. Zheng S, Black DL (2013) Alternative pre-mRNA splicing in neurons: growing up and extending its reach. Trends Genet 29:442–448
2. Chen M, Manley JL (2009) Mechanisms of alternative splicing regulation: insights from molecular and genomics approaches. Nat Rev Mol Cell Biol 10:741–754
3. Hartmann B, Valcarcel J (2009) Decrypting the genome's alternative messages. Curr Opin Cell Biol 21:377–386
4. Chen L, Zheng S (2009) Studying Alternative splicing regulatory networks through partial correlation analysis. Genome Biol 10(1):R3
5. Witten JT, Ule J (2011) Understanding splicing regulation through RNA splicing maps. Trends Genet 26:6739–6747
6. Kar A, Havlioglu N, Tarn WY, Wu JY (2006) RBM$ interacts with an intronic element and stimulates tau exon 10 inclusion. J Biol Chem 281:24479–24488
7. Wu YY, Kar A, Kuo D, Yu B, Havlioglu N (2006) SRp54 (SFR11), a regulator for tau exon 10 alternative splicing identified by an expression cloning strategy. Mol Cell Biol 26:6739–6747
8. Oberdoerffer S, Moita LF, Neems D, Freitas RP, Hacohen N, Rao A (2008) Regulation of CD45 alternative splicing by heterogenous ribonucleoprotein, hnRNPLL. Science 321:686–691
9. Topp JD, Jackson J, Melton aA, Lynch KW (2008) A cell-based screen for splicing regulators identifies hnRNP LL as a distinct signal-induced repressor of CD45 variable exon 4. RNA 14:2038–2049
10. Kuroyanagi H, Ohno G, Sakane H, Maruoka H, Hagiwara M (2010) Visualization and genetic analysis of alternative splicing regulation in vivo using fluorescence reporters in transgenic *Caenorhabditis elegans*. Nat Protoc 5:1495–1517
11. Kuroyanagi H, Kobayashi T, Mitani S, Hagiwara M (2006) Transgenic alternative splicing reporters reveal tissue-specific expression profiles and regulation mechanisms in vivo. Nat Methods 3:909–915
12. Stoilov P, Lin CH, Damoiseaux R, Nikolic J, Black DL (2008) A highthroughputscreening stragety identifies cariotonic steroids as alternative splicing modulators. Proc Natl Acad Sci 105:11218–11223
13. Warzecha CC, Sato TK, Nabet B, Hogenesch JB, Carstens RP (2009) ESRP1 and ESRP2 are epithelial cell-type specific regulators of FGFR2 splicing. Mol Cell 33:591–601
14. Moore MJ, Wang Q, Kennedy CJ, Silver PA (2010) An Alternative splicing network links cell-cycle control to apoptosis. Cell 142:625–636
15. Zheng S, Damoiseaux R, Chen L, Black DL (2013) A broadly applicable high-throughput screening strategy identifies new regulators of Dlg4 (Psd-95) alternative splicing. Genome Res 23:998–1007
16. Chen L, Zheng S (2008) Identify alternative splicing events based on position-specific evolutionary conservation. PLoS One 3(7):e2806

Chapter 6

High-Throughput Screening of a Luciferase Reporter of Gene Silencing on the Inactive X Chromosome

Alissa Keegan, Kathrin Plath, and Robert Damoiseaux

Abstract

Assays of luciferase gene activity are a sensitive and quantitative reporter system suited to high-throughput screening. We adapted a luciferase assay to a screening strategy for identifying factors that reactivate epigenetically silenced genes. This epigenetic luciferase reporter is subject to endogenous gene silencing mechanisms on the inactive X chromosome (Xi) in primary mouse cells and thus captures the multilayered nature of chromatin silencing in development. Here, we describe the optimization of an Xi-linked luciferase reactivation assay in 384-well format and adaptation of the assay for high-throughput siRNA and chemical screening. Xi-luciferase reactivation screening has applications in stem cell biology and cancer therapy. We have used the approach described here to identify chromatin-modifying proteins and to identify drug combinations that enhance the gene reactivation activity of the DNA demethylating drug 5-aza-2′-deoxycytidine.

Key words X chromosome inactivation, Chromatin, High-throughput screening, DNA methylation, Luciferase

1 Introduction

In the field of cancer biology there is a growing appreciation that tumorigenesis is frequently driven by epigenetic events such as tumor suppressor gene silencing by DNA methylation [1, 2]. Development of therapies targeting epigenetic pathways is hampered by the difficulty of monitoring treatment efficacy. Drugs such as DNA methylation inhibitors and histone deacetylase inhibitors are believed to reprogram the oncogenic cell fate through gene expression changes [2, 3]. Conventional assays of cytotoxicity such as those applied to chemotherapeutics are not ideal to measure epigenetic activity of drug candidates [3]. Changes in gene expression secondary to epigenetic reprogramming may take weeks to take effect and can be difficult to reproduce ex vivo [3]. Therefore, accurate reporter systems are needed to develop epigenetically acting drug regimens. As an example, Cui and coauthors described

Robert Damoiseaux and Samuel Hasson (eds.), *Reporter Gene Assays: Methods and Protocols*, Methods in Molecular Biology, vol. 1755, https://doi.org/10.1007/978-1-4939-7724-6_6,

such a reporter gene system that demonstrated clinically relevant behavior in response to pharmacologic treatment [4]. They targeted a fluorescent reporter downstream of an aberrantly methylated tumor suppressor gene, *SFRP1*, in a colon cancer cell line and used *SFRP1-GFP* activation to monitor the effect of the DNA demethylating drug 5-aza-2′-deoxycytidine (5-aza-2′-dC or decitabine) [4]. They found that more prolonged 5-aza-2′-dC exposure promoted *SFRP1-GFP* expression, which may help explain why many patients with acute myeloid leukemia and myelodysplastic syndrome demonstrate a delayed response to 5-aza-2′-dC treatment [4, 5]. The reporter approach described here is of similar principle but makes modifications to the reporter gene that are more compatible with high-throughput screening and that have the potential to lead to more diverse, biologically relevant findings. First, we use a luciferase reporter that replaces fluorescence imaging with a rapid, enzymatic assay that gives quantitative readout of reporter activity. Second, rather than targeting the reporter to an aberrantly methylated gene, we have used a targeting approach that leads to endogenous reporter gene silencing in mouse development through the process of X chromosome inactivation (XCI). By using XCI as the model system of gene silencing and chromatin change, we increase the potential applications beyond cancer therapy to stem cell biology, as we describe below.

We adapted the model system of XCI to siRNA and chemical screening. XCI is one of the most studied examples of gene silencing in mammals and affects an entire chromosome. It is a mechanism of dosage compensation for X-linked genes between mammalian sexes that involves chromosome-wide transcriptional silencing beginning in early female embryonic development [6]. Initiation of XCI occurs with sequential events including upregulation and spread of the long noncoding RNA *Xist*, loss of RNA Polymerase II, gain of repressive histone methylation marks such as histone H3K27me3 and H3K9me2, histone deacetylation, and gain of DNA methylation at CpG islands [7]. The choice of which of the two X chromosomes is inactivated is random, however, if *Xist* is deleted off one X chromosome in female mice, that X chromosome will remain active and the wildtype X chromosome will undergo inactivation in 100% of cells [8]. Once established, the inactive X chromosome (Xi) is remarkably robust as it is maintained for the lifetime of the female [6]. Complete X chromosome reactivation (XCR) occurs in the female germline to transmit genetic information from both X chromosomes [6]. In tissue culture, complete XCR occurs from somatic mouse cells that undergo successful transcription-factor induced reprogramming to the induced pluripotent stem cell (iPSC) state [9]. XCR occurs as one of the final events in the course iPSC reprogramming [10]. XCR is a faithful marker of complete reprogramming to the iPSC state and has immediate applications in stem cell biology. For instance,

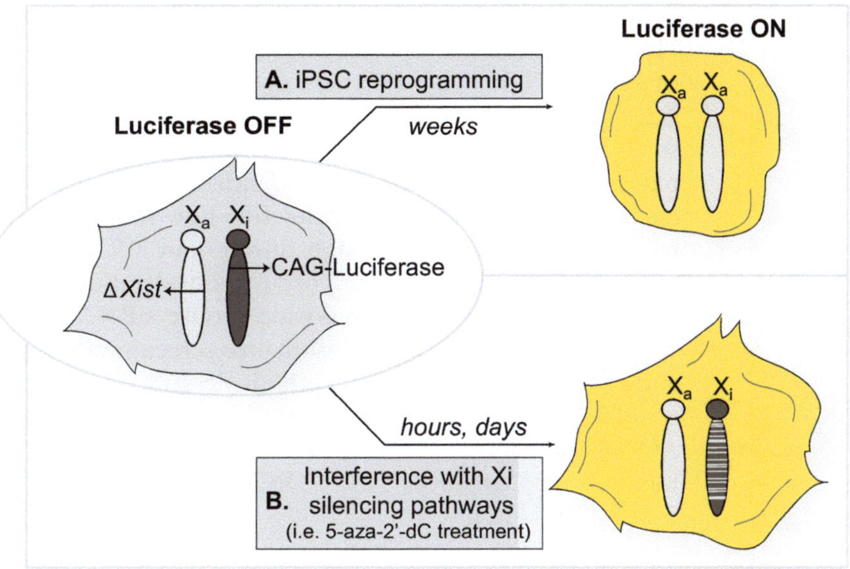

Fig. 1 Applications of X chromosome reactivation screening. Primary mouse embryonic fibroblasts (MEFs) bearing a luciferase transgene on the inactive X chromosome (Xi) do not have luciferase signal at baseline. (**a**) Transcription factor-induced reprogramming to iPSCs culminates in complete reactivation of the Xi to an active X chromosome (Xa) with concomitant luciferase (yellow) upregulation. (**b**) Interference with individual repressive chromatin pathways present on the Xi in MEFs leads to upregulation of some Xi-linked genes including the luciferase reporter in a percentage of treated cells

screens can be designed to increase the efficiency of iPSC reprogramming by monitoring for earlier and higher rates of XCR in response to protein overexpression, gene knockdown, or chemical treatment (Fig. 1).

Monitoring of XCR is also relevant to cancer biology. Partial XCR can be elicited by targeting the pathways of DNA methylation and histone deacetylation with the same epigenetic drugs used in cancer treatment such as 5-aza-2′-dC and trichostatin A [3, 11]. Partial XCR means that for a treated cell population, only a small percentage of cells reactivate the expression of inactive X chromosome (Xi)-linked gene being assayed. Partial XCR likely reflects that multiple layers of chromatin silencing with some redundancy maintain the Xi and that the threshold for reactivation differs across cells in a population and across loci on the Xi. Nonetheless, the XCR effect of various chemical and gene knockdown treatments is reproducible [11, 12]. A benefit to studying epigenetic drug targets in the XCI model system is that the activity of treatments can be detected by XCR activity in as few as 1–3 days [12]. The short timecourse of the reporter assay in primary cells avoids epigenetic drift and secondary cellular changes that may indirectly affect chromatin silencing. As we show here, XCR is highly amenable to high-throughput screening. Cells bearing a luciferase reporter gene on the Xi do not have detectable background reactivation [12]. The luciferase reporter is also sensitive and detects low rates of partial XCR [12]. Furthermore, the Xi-luciferase reporter is suitable to combination drug screening since it undergoes in vivo silencing due to XCI, which involves multiple repressive chromatin pathways, and thus reflects the multilayered nature of repressive chromatin changes. Here, we describe how to adapt an XCR assay to 384-well format with high-throughput siRNA and chemical screening. We detail the important considerations when scaling the experiment to a genome-wide mouse siRNA library. Finally we touch on approaches to normalize screening data and identify and validate hits. Using the methods described here, we were able to identify knockdown of a chromatin silencing factor *Atf7ip* as an XCI maintenance factor and to identify a combination drug treatment that enhances the DNA demethylating ability of 5-aza-2′-dC [12].

2 Materials

1. Mouse embronic fibroblast (MEF) medium, 1×: Dulbecco's modified Eagle medium supplemented with 10% fetal bovine serum, nonessential amino acids, L-glutamine, penicillin–streptomycin, and β-mercaptoethanol.
2. PLB lysis buffer (Promega).
3. Luciferase assay reagent I (LAR) (Promega).
4. GloMax microplate luminometer (Promega).
5. Genome-wide mouse siRNA library provided as Silencer Mouse Druggable Genome siRNA library V3 and Silencer Mouse Genome siRNA V3 Extension Set (Ambion, Life Technologies).
6. Multidrop 384 reagent dispenser (Thermo Scientific).
7. BenchCel 4× plate handler with Vcode bar code print and apply station, and Vprep pipetting system with a 96 LT head (all Velocity11, Agilent Technologies).
8. Matrix 384-well tissue culture plates (Thermo Scientific).
9. PlateLoc thermal microplate sealer (Agilent Technologies).
10. Dnmtl siRNA (Ambion, ID # AM161526).
11. 5-Aza-2′-deoxycytidine (5-aza-2′-dC) (Sigma-Aldrich). Resuspend in DMSO and bring to 100 mM, then freeze aliquots in −80 °C. Thaw aliquot at room temperature just before adding to tissue culture media.

12. Microsource, Biomol enzyme inhibitor and bioactive lipid libraries, Prestwick chemical library, and NIH clinical collections. Stored at 10 mM in DMSO.
13. 2× MEF medium: Dulbecco's modified Eagle medium supplemented with 20% fetal bovine serum, 2× nonessential amino acids, 2× L-glutamine, 2× penicillin–streptomycin, and 1× β-mercaptoethanol.
14. 5× siRNA buffer (Dharmacon, GE Healthcare).
15. Tissue-culture grade phosphate buffer saline (PBS) (Life Technologies).
16. Trypsin–EDTA 0.25%, phenol red (Life Technologies).
17. Opti-MEM reduced serum medium (Life Technologies).
18. Lipofectamine RNAiMAX transfection reagent (Life Technologies).
19. ELx405 select deep well washer (Bio-Tek Instruments).
20. ONE-Glo luciferase assay reagent (Promega).
21. FlexStation II benchtop multimode microplate reader (Molecular Devices).
22. Biomek FXP laboratory automation workstation (Beckman Coulter).

3 Methods

3.1 Preparing Xi-Luciferase Reporter Cells

MEFs were isolated from Xi-luciferase female transgenic embryos bearing a CAG promoter-driven luciferase transgene in the *Hprt* locus on the X chromosome as well as deletion of *Xist* on the other X chromosome [12]. The deletion of *Xist*, a noncoding RNA required for XCI, ensures that the luciferase-bearing transgene is silenced in all cells [8, 12]. Xi-luciferase female double transgenic embryos ($Xi^{CAG\text{-}Luciferase}Xa^{\Delta Xist}$) are derived at an expected 1:4 Mendelian ratio from a cross of male luciferase reporter mice ($X^{CAG\text{-}Luciferase}Y$) and female mice heterozygous for *Xist* knockout ($X^{\Delta Xist}X^{wt}$). This cross also produces littermate females lacking the *Xist* deletion ($X^{CAG\text{-}Luciferase}X^{wt}$) that express luciferase at baseline due to random XCI. Therefore, genotyping and careful tissue culture technique are necessary to avoid contamination with luciferase-expressing littermate cells.

1. Isolate MEFs by standard methods from embryos (d13 or d14 d.p.c.) in 1× MEF media in 15 cm^2 plates (*see* **Note 1**) [13].
2. Upon reaching 80–90% confluence, sample MEFs for *Hprt*, *Xist*, and *Zfy* (to confirm absence of Y chromosome) genotyping by PCR, and freeze the remainder of cells in 1× MEF media with 20% DMSO.

3. If MEFs are identified from the litter as Xi-luciferase MEFs on the basis of carrying both the luciferase allele and the *Xist* deletion, thaw the Xi reporter MEFs in 15 cm^2 plates and passage two more times at a 1:6 split. With the first passage, a sample of cells should be tested to confirm absence of luciferase activity at baseline (*see* **Note 2**). Plate a suspension of 60,000 cells in a 12-well plate well for 48–72 h, then lyse the adherent cells in 200 μL of PLB buffer for 20 min on an orbital shaker. Clear the lysate with a short centrifugation step of 12,000 × *g* for 30 s then assay 20 μL of cleared lysate with 50 μL of LARI reagent on a luminometer. At the second split, MEFs derived from different Xi-luciferase reporter embryos should be pooled together to ensure homogeneity in the batches frozen for the screen. Freeze each 15 cm^2 plate of cells to one vial. Thaw one vial to two 15 cm^2 plates and count cells 24 h later to estimate yield from the frozen batches in preparation for screening (*see* **Note 3**).
4. In addition to thawing cells 24 h prior to screening, autoclave one tissue-culture grade 500 mL bottle with a magnetic stir bar inside per each set of 30 library plates that will be screened. This glassware will be used to sterilely dispense a homogenous suspension of cells to 384-well plates.

3.2 siRNA Library Preparation

We chose to use a genome-wide mouse siRNA library provided as 0.25 nmol dried oligonucleotide in 153 384-well plates. Two columns were left empty on each plate for positive siRNA control and no siRNA control. The library was resuspended by centrifuging each plate for 2 min at 1000 × *g*, adding 50 μL of RNAse free water with a Multidrop reagent dispenser in an RNAse-free biological safety cabinet, resealing, vortexing, and centrifuging again at 1000 × *g* for 2 min. Four sample plates were randomly chosen to confirm RNA concentration between 64 and 75 ng/μL. Copies of the library were transferred by BenchCel 4× plate handler with Vcode bar code application, and Vprep pipettor with a 96 LT head. 2 μL of siRNA suspension from source plates were moved into white opaque 384-well tissue culture plates, sealed with plate sealer, and then frozen in −80 °C.

3.3 Screen Controls

Loss of DNA methylation by deletion of or interference with the maintenance methyltransferase *Dnmt1* is known to elicit XCR. Therefore we tested combinations of *Dnmt1* knockdown and 5-aza-2′-dC, a DNMT1-inhibiting chemical, treatment to produce luciferase signal in 384-well format [11, 14]. We chose a *Dnmt1* siRNA that reduced *Dnmt1* RNA levels to <10% of control levels as measured by RT-qPCR from MEFs treated in the 12-well format [12]. We then moved to optimizing the assay in 384-well format and found that *Dnmt1* knockdown in the presence of low concentration 5-aza-2′-dC (0.2 μM), reliably produced luciferase reactivation [15]. Addition of 5-aza-2′-dC was likely necessary to reduce DNA methylation to a lower level such that knockdown of *Dnmt1* boosted Xi-luciferase reactivation signal to a threshold detectable in the 384-well assay. The low concentration of 5-aza-2′dC without *Dnmt1* knockdown did not produce background signal [15]. Therefore, we added siRNA against *Dnmt1* to all 16 wells of the last column of each siRNA screening plate as a positive control. We also added 5-aza-2′-dC to a final concentration of 0.2 μM to each screening well including positive and negative controls and library samples. The negative control, occupying the 16 wells in the penultimate column of each screening plate, contained all the reagents with exception of siRNA. For the chemical screen, we used high concentration 5-aza-2′-dC (10 μM) in one row of each plate as a positive control.

3.4 siRNA Knockdown Assay Optimization

The optimization of siRNA knockdown in 384-well format is assessed by reporter activity with positive control knockdown. As with knockdown of *Dnmt1* (*see* Subheading 3.3), this siRNA should be validated for target gene knockdown from a larger assay format such that RNA yield is sufficient for RT-qPCR. Once a positive control such as the *Dnmt1* knockdown and 5-aza-2′-dC (0.2 μM) treatment is established (*see* Subheading 3.3), the assay can be optimized to 384-well format by comparing the signal distribution between positive and negative control samples. We recommend optimizing the assay to maximize the *Z*-factor, which is a coefficient reflective of assay dynamic range and data signal variations (Fig. 2) [16]. A feasible assay for screening has a *Z*-factor > 0, and an excellent assay has a *Z*-factor higher than 0.5. Pilot experiments should sequentially test experimental variables for raising the *Z*-factor of the assay (*see* **Note 4**). The experimental variables in our optimization are summarized in Table 1 in the order that we tested them (*see* **Note 5**). Since our final assay *Z*-factor was 0.11, which is predictive of generating false positives and false negatives, we screened the siRNA library in duplicate to add statistical power.

3.5 siRNA Screening of Xi-Reporter MEFs

1. One day prior to transfection thaw and plate the Xi-luciferase reporter MEFs. The number of vials of cells should be sufficient to distribute 2000 cells to each 384-well well 24 h later, accounting for 20% extra cells (*see* **Note 6**). We recommend screening one 30-plate batch from the library on the first day of screening then increasing to two 30-plate batches on subsequent days. The determination of the number of vials of cells to thaw should be made from a previous test of cell yield from the frozen batches (refer to Subheading 3.1).

$$\text{Z-factor} = 1 - \frac{3\,(\text{StDev}_{\text{sample}}) + 3\,(\text{StDev}_{\text{control}})}{|\text{Mean}_{\text{sample}} - \text{Mean}_{\text{control}}|}$$

Adapted from Zhang et al., 1999

Fig. 2 *Z*-factor quality coefficient for screening assay

Table 1
Experimental variables for screening optimization

Experimental variable	Optimal condition	Other condition(s) assayed
Cell Number	2000 cells	750–2500 cells
5-aza-2′-dC concentration	0.2 μM	0.05–0.15 μM
Incubation time after knockdown	72 h	48 or 60 h
Luciferase plate reader	Acquest (Molecular Devices)	Wallac 1420 Victor2 (Perkin Elmer) or Tecan
384-well plate	Matrix (Thermo Scientific)	Greiner (Sigma-Aldrich)
Luciferase assay system	ONE-Glo (Promega)	Bright-Glo (Promega)
Luciferase reagent volume	20 μL	10–30 μL
Transfection reagent type	RNAimax (Life Technologies)	DharmaFECT (Dharmacon)
Transfection reagent volume	0.03 μL	0.02–0.15 μL
Luciferase assay incubation time	20 min	45 min
Culture media aspiration step	Incude	Exclude

2. Prepare 2× MEF media. This 2× MEF media will be diluted in each sample by reduced-serum media in the transfection step.
3. Thaw the 30 siRNA screening plates per batch, centrifuge for 1 min at 1000 × *g*, wipe down with RNAse reducing solution in a Biological Safety Cabinet, peel off cover, and stack screening plates.
4. Dilute *Dnmt1* siRNA in 5× siRNA Buffer to distribute 2 μL solution containing 1 pmol siRNA to the 16 wells of the positive control column.
5. Trypsinize MEFs thawed onto 15 cm^2 plates 24 h prior by washing in PBS then treating for 5 min with 2.5 mL of Trypsin-EDTA 0.25% in a humidified 37 °C incubator. Resuspend trypsinized cells in 18 mL of MEF media and pool into 50 mL tissue-grade conical tubes. Centrifuge the cells at

300 × *g* for 5 min, then carefully aspirate off supernatant, and store cell pellet on ice.

6. Dilute a stock of 5-aza-2′-dC from 100 mM in DMSO to 100 μM in 2× MEF media.
7. Mix a transfection solution of Opti-MEM reduced serum media at room temperature and Lipofectamine RNAimax transfection mix in a ratio of 20 μL of Opti-MEM to 0.03 μL RNAimax per each well, again accounting for 20% extra solution. Incubate the transfection solution for 20 min prior to adding to siRNA.
8. Using a Multidrop, distribute 20 μL of transfection mix to each well of the plates including positive controls and the no siRNA column. Ensure transfection mix incubates with siRNAs from 30 min to 1 h prior to addition of cells.
9. Assuming 20 μL of cell suspension is administered to each well, depending on the number of 384 plates being screened, transfer the appropriate volume of 2× MEF media warmed to 37 °C into the 500 mL tissue culture bottle with stir bar. Resuspend the cell pellet with 10 mL of this 2× MEF media and transfer to the 500 mL bottle, carefully triturating to homogenize the cells and agitating on a stir plate. Dilute 5-aza-2′-dC to 0.04 μM in this cell solution. Deliver 20 μL of the cell and 5-aza-2′-dC solution to the transfection mix while maintaining gentle stirring to avoid cell clumping (*see* **Note 7**).
10. Incubate cells in siRNA knockdown mixture with 5-aza-2′-dC for 72 h in a humidified 37 °C incubator at 5% CO_2.
11. In batches of 14 plates, prior to measuring luciferase activity, aspirate off 20 μL of media using a well washer (*see* **Note 8**).
12. Using a Multidrop distribute 20 μL of One-Glo luciferase assay reagent and incubate for 20 min.
13. Measure luciferase activity with microplate reader. Perform visual inspection of plate heat map for presence and absence of high luciferase signal in positive and negative control wells, respectively. Take note of any plates that have a higher density of high luciferase values as they make represent technical errors.

3.6 Chemical Screening of Xi-Reporter MEFs

1. Analogously to the siRNA screening protocol, 1 day prior to treatments, thaw approximately two vials of cells in 4 15 cm^2 plates, or enough cells to distribute 2000 cells to each 384-well 24 h later, accounting for 20% extra.
2. Prepare 50 μL of cell suspension with 2000 cells per well in 1× MEF media with 0.2 μM 5-aza-2′-dC, accounting for 20% extra. As before, prepare cell suspension in a 500 mL bottle with magnetic stir bar while distributing 50 μL of solution per well by Multidrop.

3. Prepare a positive control mixture for a row of wells on each plate by mixing 50 μL of cell suspension with 2000 cells per well in 1× MEF media with high concentration 5-aza-2′-dC (10.0 μM). Distribute 50 μL of positive control mixture by multichannel pipette.
4. Add 0.5 μL of screening chemical in DMSO by liquid handling system to rows excluding the positive control row.
5. Incubate cells in chemical treatment mixture with 5-aza-2′-dC for 72 h in a humidified 37 °C incubator at 5% CO_2.
6. As with siRNA screening approach, measure luciferase activity with microplate reader with visual inspection of plate signal heat map.

3.7 Data Analysis

3.7.1 Data Normalization

High-throughput screening with siRNAs, as opposed to chemicals, is subject to unique sources of variability partly due the biology of siRNA with off-target effects and partly due to the required addition of a transfection step [17]. Systematic error can be seen within a batch as incubation times vary across plates or with regard to plate layout as siRNA screens are more susceptible to edge effects [17]. The first level of quality assurance should be in real-time with visualization of plate luciferase values as they are collected. Many plate readers, such as the FlexStation microplate reader, display a heat map with real-time data collection, which facilitates identification of errors so that plates can be tagged for closer analysis later and so that technical errors can be immediately corrected. Data normalization can be achieved by comparing the sample to positive or negative controls or by comparing to the individual plate distribution as part of the *z*-score (*see* **Note 9**) [17]. *See* Birmingham et al. [17] for an in-depth discussion of data normalization strategies.

3.7.2 Hit Selection

The identification of hits from siRNA library screens is complicated by a high rate of false positives [17]. The main feature that enhances selection of biologically active hits from siRNA screens is the redundancy included in the siRNA library design [17]. For instance, the genome-wide mouse Silencer siRNA library includes three unique siRNAs against each gene. We increased redundancy by screening this library in duplicate because of the mediocre *Z*-factor of our optimized assay. We pooled data such that six data points represented each gene. Then we applied the Redundant siRNA Activity (RSA) analysis which ranks all siRNAs by normalized value (robust *z*-score in this case) then analyzes siRNAs according to their gene targets [18]. Genes are assigned *p*-values on the basis of whether siRNAs against that individual gene cluster higher in the ranking than would be expected by chance [18]. Validation of the method of analysis can be confirmed by identification of the gene target of the positive control siRNA from the library by RSA analysis. For instance, our top RSA hit with the *p* value of 3.2×10^{-6} was *Dnmt1* [15].

3.7.3 Hit Validation

The next step in data analysis is to generate a list of top-scoring hits for subsequent validation. For a high quality screen with a *Z*-factor ≥ 0.5 one may arbitrarily chose a cutoff by considering a technically feasible number of siRNAs to validate. Alternatively, one may hand-annotate gene function or apply gene ontogeny analysis to decide on a cutoff. In the case of the Xi-luciferase screen, the 50 top genes by RSA analysis (with the lowest RSA *p*-value) included genes with known chromatin functions but also genes that were more likely to represent false positives such as olfactory receptor genes and transmembrane channel subunits. Thus, we chose a cutoff of the top 54 genes but omitted those genes that were unlikely to have a role in Xi chromatin maintenance (*see* **Note 9**). For the validation assays, we recommend ordering resynthesized active siRNAs against the gene hits in order to ensure that the library annotation is accurate. To complete the validation of active siRNAs against gene hits, it is important to rule out a luciferase-specific effect. We retested active siRNAs with Xi-H2B-Citrine reporter MEFs that were synonymous to the Xi-luciferase reporter MEFs but carry a fluorescent reporter in the place of luciferase [12]. In summary, data analysis includes choosing an appropriate method to normalize data, applying statistical techniques to rank genes, determining a cutoff for hits, and validating the hit siRNA sequences in an assay with a different reporter gene readout. For chemical screening, many of these same considerations in data analysis and hit confirmation apply [19]. We performed both chemical and siRNA screening with the Xi-luciferase assay, as described here, and identified a protein target common to an siRNA and a chemical hit [15]. This overlap suggested identification of a biologically relevant pathway and thus we prioritized validation of these hits. We found the two screening approaches to be highly complimentary; the siRNA hit helped identify the intracellular target of the chemical compound and the identification of chemical compound allowed us to translate our observations to a disease model in which the chemical was used as a drug [15]. We believe that redundancy is a major characteristic of screening design that ensures identification of high confidence hits. Redundancy can be accomplished by using a library with many reagents directed against common pathways, by testing those reagents in replicates, and by screening with multiple modalities such as with siRNA and chemical libraries.

4 Notes

1. We found that MEFs derived from four individual Xi-luciferase embryos were sufficient to screen 300 384-well plates.
2. Measuring luciferase activity in addition to checking genotypes is an important control to avoid contaminating Xi-reporter cell stocks with luciferase-expressing littermate cells.
3. In our experience using 2000 cells per 384-well screening assay, we could screen 30 library plates/day by thawing two vials of cells to 4 15 cm^2 plates 24 h prior.
4. The methods describe a screen using empirically determined "optimal variables" as listed in Table 1. With each screen, we recommend optimizing to variables such as these using the *Z*-factor as shown in Fig. 2 to measure assay quality.
5. We recommend using at least half of a 384-well plate per variable to better assess the distribution of signal with each experimental condition. While early phases of optimization can include manually pipetted reagents, later optimization trials should attempt to replicate actual screening conditions including the use of manifold liquid handlers and other automated steps.
6. Whenever making large volumes of solutions to be aliquoted by Multidrop or multichannel pipette, calculate for 20% of extra solution considering losses in manifold tubing or reservoir.
7. Gentle agitation of the cell suspension is an important step for homogenous distribution of healthy cells. Avoid media foaming.
8. We found this step of removing media from adherent MEF cultures to have a large effect on increasing luminescence signal. Removing the supernatant increases the concentration of luciferase in the cell lysate in the next step.
9. An important caveat to using individual plate distribution for normalization is that the screening library has a relatively equal distribution of hits across plates due to random organization of genes in the library. We chose to normalize to a robust *z*-score which takes into account plate median and median absolute deviation and is therefore less sensitive to outliers than *z*-score [17].
10. Hand annotation should be performed with caution because it introduces bias and limits further exploration of unexpected and interesting gene pathways.

Acknowledgments

This work was supported National Research Service Award AG039179 to A.K. and by funds from the Iris Cantor-UCLA Women's Health Center Executive Advisory Board to K.P./A.K. K.P. is supported by the NIH (DP2OD001686 and P01 GM099134), CIRM (RN1-00564, RB3-05080, and RB4-06133), and the Jonsson Comprehensive Cancer Center and the Eli and Edythe Broad Center of Regenerative Medicine and Stem Cell Research at UCLA. We are grateful to Winnie Hwong for technical assistance, and to Stephen T. Smale for experimental advice.

References

1. Esteller M (2008) Epigenetics in cancer. N Engl J Med 358:1148–1159
2. Jones PA, Baylin SB (2007) The epigenomics of cancer. Cell 128:683–692
3. Azad N, Zahnow CA, Rudin CM, Baylin SB (2013) The future of epigenetic therapy in solid tumours--lessons from the past. Nat Rev Clin Oncol 10:256–266
4. Cui Y et al (2014) A recombinant reporter system for monitoring reactivation of an endogenously DNA hypermethylated gene. Cancer Res 74:3834–3843
5. Issa J-PJ, Kantarjian HM (2009) Targeting DNA methylation. Clin Cancer Res 15:3938–3946
6. Lee JT (2011) Gracefully ageing at 50, - X-chromosome inactivation becomes a paradigm for RNA and chromatin control. Nat Rev Mol Cell Biol 12:815–826
7. Chow J, Heard E (2009) X inactivation and the complexities of silencing a sex chromosome. Curr Opin Cell Biol 21:359–366
8. Marahrens Y, Panning B, Dausman J, Strauss W, Jaenisch R (1997) Xist-deficient mice are defective in dosage compensation but not spermatogenesis. Genes Dev 11:156–166
9. Maherali N et al (2007) Directly reprogrammed fibroblasts show global epigenetic remodeling and widespread tissue contribution. Cell Stem Cell 1:55–70
10. Pasque V et al (2014) X chromosome reactivation dynamics reveal stages of reprogramming to pluripotency. Cell 159:1681–1697
11. Csankovszki G, Nagy A, Jaenisch R (2001) Synergism of Xist RNA, DNA methylation, and histone hypoacetylation in maintaining X chromosome inactivation. J Cell Biol 153:773–784
12. Minkovsky A et al (2014) The Mbd1-Atf7ip-Setdb1 pathway contributes to the maintenance of X chromosome inactivation. Epigenetics Chromatin 7:12
13. Jozefczuk J, Drews K, Adjaye J (2012) Preparation of mouse embryonic fibroblast cells suitable for culturing human embryonic and induced pluripotent stem cells. J Vis Exp 64
14. Sado T et al (2000) X inactivation in the mouse embryo deficient for Dnmt1: distinct effect of hypomethylation on imprinted and random X inactivation. Dev Biol 225:294–303
15. Minkovsky A et al (2015) A high-throughput screen of inactive X chromosome reactivation identifies the enhancement of DNA demethylation by 5-aza-2'-dC upon inhibition of ribonucleotide reductase. Epigenetics Chromatin 8:42
16. Zhang JH, Chung TD, Oldenburg KR (1999) A simple statistical parameter for use in evaluation and validation of high throughput screening assays. J Biomol Screen 4:67–73
17. Birmingham A et al (2009) Statistical methods for analysis of high-throughput RNA interference screens. Nat Methods 6:569–575
18. Konig R et al (2007) A probability-based approach for the analysis of large-scale RNAi screens. Nat Methods 4:847–849
19. Malo N, Hanley JA, Cerquozzi S, Pelletier J, Nadon R (2006) Statistical practice in high-throughput screening data analysis. Nat Biotechnol 24:167–175

Chapter 7

Making It All Work: Functional Genomics and Reporter Gene Assays

Genevieve Welch, Robert Damoiseaux, and Loren Miraglia

Abstract

Functional genomics is the study of the function of genes on a genome-wide level. Reporter gene assays can be utilized in this context to dissect signaling cascades, find new drug targets, or decipher the function of gene expression. The genome-wide scale of these experiments necessitates a different approach toward science than traditional single hypothesis driven research. High-throughput experimentation requires large project teams, automation, and discrete validation of each step in the automation and assay process. The purpose of this chapter is to provide a general outline of a standard functional genomics project with a reporter gene assay as readout, give an overview of the methodologies employed and familiarize the reader with the subsequent data analysis. The advantages of such high throughput experimentation are speed, quantitative results, and insights into biology on a genome-wide scale all of which enable a more rapid progress of science.

Key words High-throughput screening, Functional genomics, Reporter gene assays, siRNA, RNAi, cDNA, shRNA, HTS, Automation, Screening, Liquid handling, Project management, Assay development, Data mining

1 Introduction

Science in the traditional laboratory has—by and large—to still take advantage of the fruits of the industrial revolution. This is mainly due to the fact that the habit of performing experiments in a one-off hypothesis driven approach is extremely common. With the completion of the human genome and the annotation of sequence with genes—at times putative and at times known—the picture has changed. While hypothesis driven approaches are still extremely useful and part of the scientific method, it has become clear that our understanding of the function of genes is lacking which inhibits the effectiveness of this approach. In fact, a conservative estimate would be that about 50% of all genes are incorrectly annotated. The situation varies between species but as the annotation was originally based on homology modeling based on the

Robert Damoiseaux and Samuel Hasson (eds.), *Reporter Gene Assays: Methods and Protocols*, Methods in Molecular Biology, vol. 1755, https://doi.org/10.1007/978-1-4939-7724-6_7, © Springer Science+Business Media, LLC, part of Springer Nature 2018

E. coli genome, the limitations are apparent. The use of reporter genes in the context of functional genomics offers a different approach to science that enables precision discovery that can be tailored to the scientific question and can range in scope from single genes over pathways to an interrogation of all genes of the genome.

Of course, with larger scope, the strategies employed change. This chapter aims at giving an introduction to large scale project approaches, lines out strategies for functional genomics screens using reporter genes and also outlines general strategies for the management of project teams. Also, these types of experiments also require a very different level of attention to detail than traditional research because the results are not a single answer to a single question but rather an ensemble of answers to a question that is interpreted by the biological system in as many ways as the experimental setup will allow. "You get what you screen for" is an old adage in high throughput screening circles and it has held true since the start of high throughput methodologies.

Specifically, this section will present an in depth overview of functional profiling or screening of functional genomics libraries using reporter gene assays: We will discuss libraries and approaches that cover arrayed cDNAs, siRNAs, miRNAs and shRNAs. These strategies apply in the same way to other screening tools such as arrayed CRISPR libraries as the optimization, employment, and analysis of any arrayed functional genomics screen are relatively similar. The general outline of a functional genomics project is displayed in Fig. 1. The main difference to most scientific projects using non-high-throughput methodologies is that the number of people involved is increased because such projects require a variety of skill sets that are unlikely to be found in a single person. Also, in addition to the people performing the work, the person or group paying for the actual work is frequently different from the people performing the work. These stakeholders need to be involved and consulted throughout the project lifetime to ensure that the execution of the project is aligned with the question that the stakeholders interested in. The correct and precise formulation of the question or goal that the screen is to answer is extremely important as is the selection of the actual experimental system. The validation of that system using proper controls and last but not least the analysis of the data and their interpretation is key to the success of the project. This requires people with not only biological skills but also people with data analysis skills in order to extract the maximal benefit from the data obtained. Due to the varied interests and skills of the different groups that participate in such a project, it is also advisable to have a project manager who interfaces will all of the parties and it is not uncommon to have one project manager for each stakeholder group.

Experimentally, this type of work is more demanding than traditional benchwork as a single failpoint will bring down the entire experiment composed of thousands of samples. In traditional work, this is not really a problem because normally the cost of failure is not detrimental but in functional genomics, the cost frequently makes failure a nonoption. This in turn requires a much more thorough and thoughtful approach toward the entire screening process and this chapter will lay out the ins and outs that govern the assay validation, execution, and data interpretation.

In summary, this comprehensive approach toward functional genomics screening of reporter gene assays is the industry standard and has the advantage that the resulting data are typically of very high quality, repeat experiments are avoided and the time needed to execute the experiments is as short as possible. Figure 1 gives an overall picture of how such large scale projects are set up and executed.

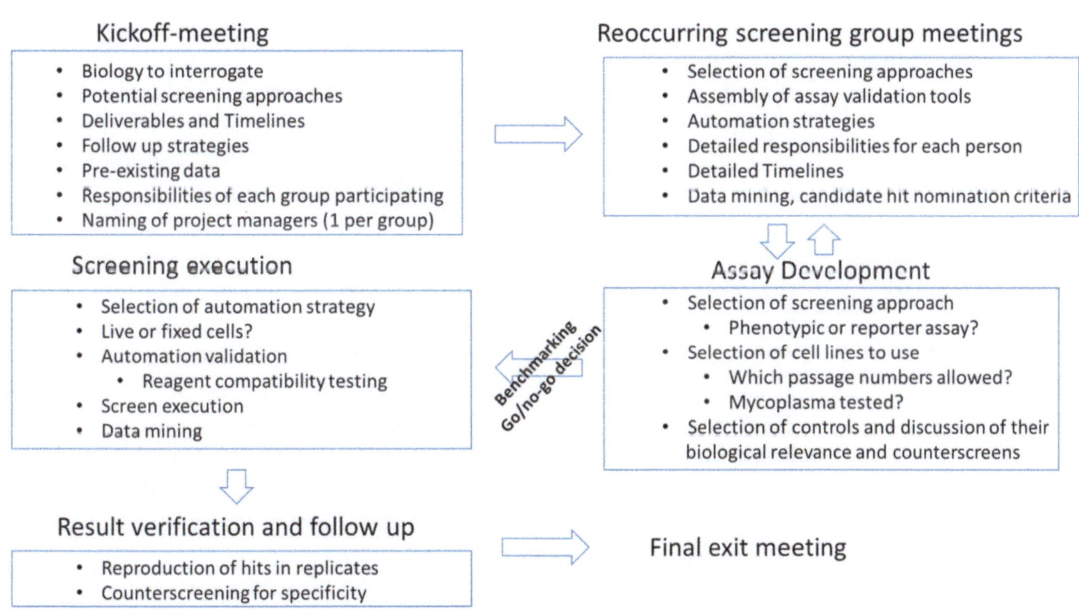

Fig. 1 General outline of the life span of a functional genomics project. The success of any large scale project is very dependent on project management efforts—without proper project management, even the best biology and equipment are ultimately not terribly useful. Hence, at the different stages of the project, it is important to formalize how the project is handled, how exchange of information is performed and roles and responsibilities are assigned. Moreover, it is important to cover the here presented topics in order to be able to measure the success of the project against hard metrics and thus allow for goals to be reached. Also, key details of topics during assay development and execution should be predetermined as outlined above

2 Materials

2.1 Cell Line

The cell line to be used in the screen needs to be purchased or generated at a low passage number and sufficient aliquots need to be banked (*see* **Notes 1–3**). Reporter genes are introduced using standard techniques. A more in depth discussion of the necessary properties of the cell line(s) employed is in Subheading 3.

2.2 Assay Reagents

All necessary assay reagents should be validated, purchased and stocked from a single lot as much as possible.

2.3 Assay Plates

Typically functional genomics assays are run in 384 well plates such as Greiner 384 well plates. Buy plates from a single lot and track to assess if the lot changes.

2.4 Function Genomics Library

The functional genomics library will typically be prestamped into 384 well plates so it can be produced before the actual assay starts. Wells for controls are left blank so it is customizable.

2.5 Detection System

The detection system employed for a functional genomics project is matched to the type of reporter gene utilized. Typically, this will be a plate reader for, for example, luminescence such as the Envision by PerkinElmer or for example an automated microscope such as the ImageXpress platform from Molecular Devices.

2.6 Bulk Dispensers and Plate Washers

Functional genomics projects take advantage of the segregation of the production of libraries and the assay execution. In order for that to be feasible, noncontact bulk dispensers are utilized to add for example reagents like transfection reagent or cells to plates that already contain the libraries. Frequently bulk dispensers such as a Multidrop Combi from Thermo or other bulk dispensers will perform fine. If needed, plate washers that can also function as bulk dispensers can be employed—e.g., the BioTek ELX 406. What is important is that these devices are heavily tested for reliability before use and maintained well.

2.7 Robotics and Robotic Scheduler

The genome-wide scale of the experiments requires the utilization of robotics and schedulers that enable the execution of work while keeping the timing of each plate within the allowable parameters of the assay. The reliability of the robotics and the scheduler are of utmost importance. Many commercial platforms exist, such as the Spinnaker robot with Momentum Scheduler from Thermo. If at all possible, utilize a preemptive scheduler, i.e., a scheduler that computes an optimal schedule and then fits the execution of the experiments around this schedule. Dynamic schedulers are useful as well but unexpected surprises do occur when trying to scale experiments and as such dummy runs without reagents are suggested.

2.8 Data Mining Software

Many statistical programs exist that can be used. Utilize a program you are comfortable with. For example the redundant siRNA analysis package is a free program for siRNA screens [1]. Other programs such as R or commercial platforms will work as well. If you are not comfortable with the data analysis part of the project, consult a biostatistician before beginning the project.

3 Methods

3.1 Project Goals and Initial Group Meeting

- Kick-off Meeting: Before beginning a screen, it is important that all stakeholders and collaborators agree—preferably in writing—to the overall goal and milestones of the project. Therefore, set up a meeting will all stakeholders and distribute minutes that outline the decision made on items below (*see* Fig. 1).
- Make a decision on the project goal. The goal definition is to include if for example a drug target, pathway or drug mode of action is to be investigated in the course of the project.
- Define the project scope: Which species and how many different cell lines and which library types are to be run. This is extremely important to ensure the maximal biological relevance of the experiments.
- Set responsibilities for specific tasks and lay out their timeline expectations.
- Define a success criteria and follow-up strategy for positive data.
- Discuss the risks of the project: This frequently covers the limitations of reagents and equipment and also entails the assessment and implementation of strategies for averting these risks.
- Discuss and assure that the biological system for the screen is relevant before committing more resources to further research or publishing the data. Biological relevance typically dictates assay parameters such as reagents and cell line.
- Review the criteria set in this initial meeting periodically to ensure that the project stays on track.

3.2 Assay Development

The general outline of the assay development is given in Fig. 2 and is mostly self-explanatory. However, several factors bear an in depth discussion as they are frequently sources of errors and problems down the road.

- Select a suitable cell line. The decision on the cell line used is likely the most important decision in functional genomics: The key is to identify a cell line that is biologically relevant. This includes markers for pathways in question, key phenotypes and also species. For example, if the siRNA, shRNA or miRNA library is designed to target human transcripts, it would not be

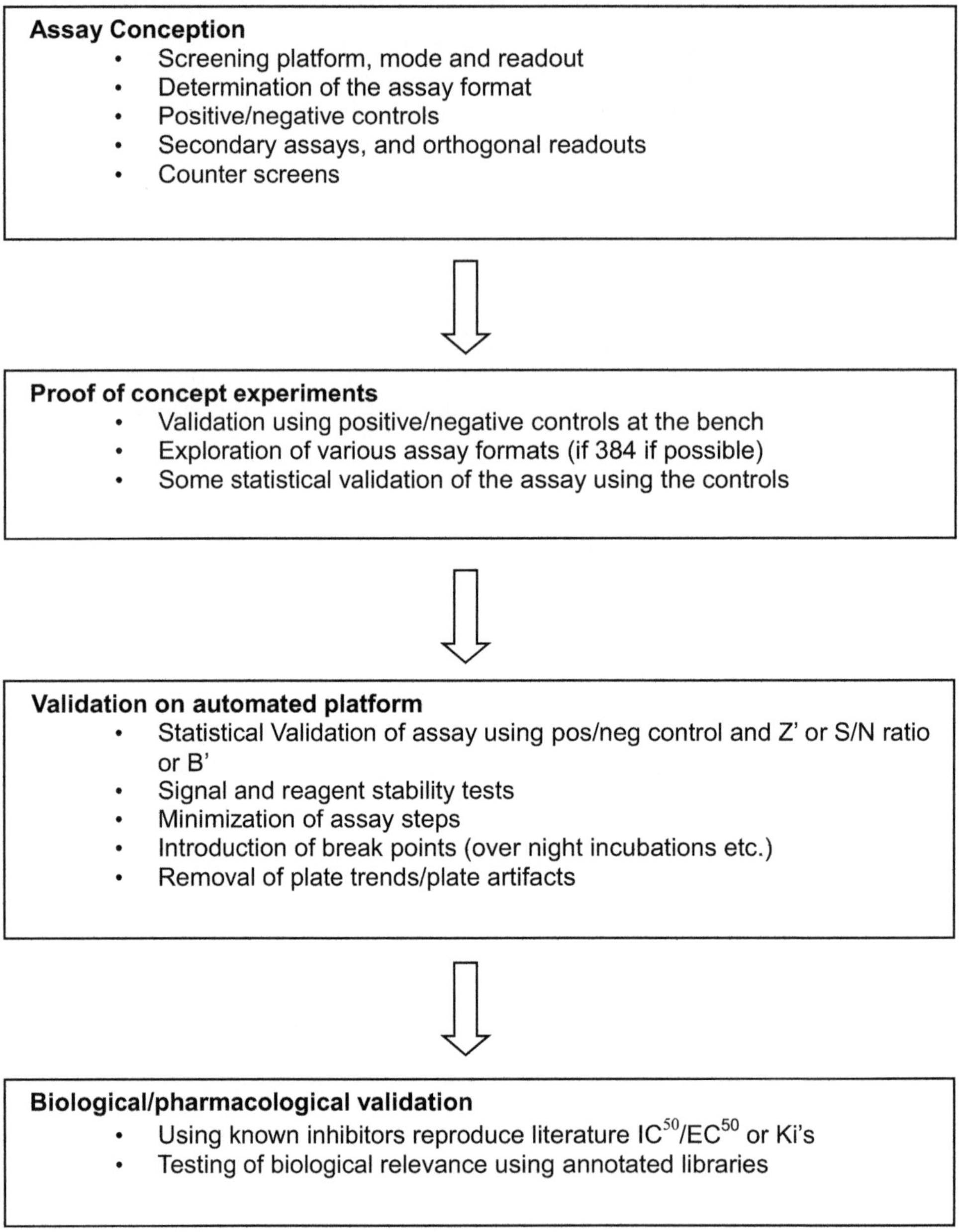

Assay Conception
- Screening platform, mode and readout
- Determination of the assay format
- Positive/negative controls
- Secondary assays, and orthogonal readouts
- Counter screens

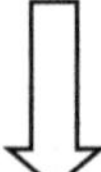

Proof of concept experiments
- Validation using positive/negative controls at the bench
- Exploration of various assay formats (if 384 if possible)
- Some statistical validation of the assay using the controls

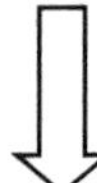

Validation on automated platform
- Statistical Validation of assay using pos/neg control and Z' or S/N ratio or B'
- Signal and reagent stability tests
- Minimization of assay steps
- Introduction of break points (over night incubations etc.)
- Removal of plate trends/plate artifacts

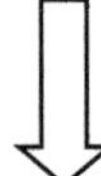

Biological/pharmacological validation
- Using known inhibitors reproduce literature IC^{50}/EC^{50} or Ki's
- Testing of biological relevance using annotated libraries

Fig. 2 Typical assay development workflow

appropriate to use a mouse line. Select a cell line species that is compatible with the functional genomics reagents that will be used.

- Validate the presence of the target/pathway under interrogation: If a particular target is not active in the interrogated cell line, then it will be difficult to find relevant hits, especially when a knock-down library is employed. This can be determined by confirming that key phenotypes or expression of pathway markers are present.

- Determine usability of the cell line for screening. Consider whether the cells can be acquired or expanded to the amounts required for screening and follow-up experiment without compromising their biological integrity. The passage number or range of passage numbers to be used during the screen must also be determined because it depends on the age of the frozen stock, the density with which it is passaged, whether the cells differentiate over passages (*see* **Note 4**).
- Determine whether the cell line requires licensing or other legal agreements. Failure to address this issue could result in the inability to use or publish the data or possible lawsuits.
- Determine the optimal delivery route for the functional genomics reagent/library. Many cell lines can be transfected and transfection is the default for all functional genomics projects. There are many transfection reagents and protocols available today and even when it seems unlikely that transfection will not work, it can often be optimized. If not, viral delivery or electroporation can be considered. (*See* **Note 5** for details.)
- Optimize the screening timeline. Figure 3 gives an overview of the various reagents and the differences in their timelines. It is also necessary to understand the biology of the pathway under investigation to make decisions regarding the screen's protocol and timeline. When possible, positive controls should be used to test different timelines and the protocol with the largest window of effectiveness chosen. Proper negative and positive controls are the corner stones of assay development (*see* **Notes 6–8**).

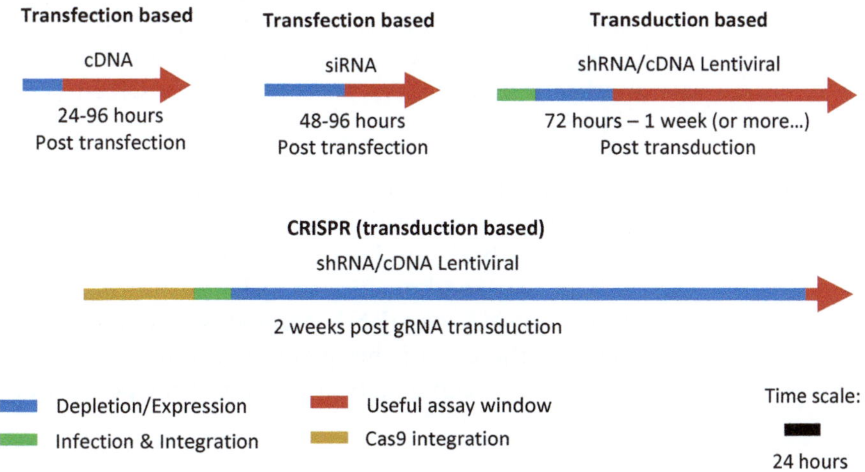

Fig. 3 Various functional genomics systems and their approximate timelines

Moreover, the controls are necessary for normalizing experimental values in order to score actives and determine outliers. More methods of normalization will be discussed in the analysis section.

- In order to optimize the cell seeding density for a screen, use the controls selected above under the same screening conditions to test their phenotypes using several cell densities. Once optimized, it is important to control tightly for the cell density (*see* **Note 9**).

3.3 Reagent and Test Set Performance

- After optimizing the screening protocol, determine how much reagent will be needed for all subsequent screens including reconfirmation screens.
- Estimate the amount of time needed to perform each step in the screen protocol as it will be performed. Using this information, test stability and performance of each reagent when prepared and waiting to be dispensed, imaged or otherwise used. Test all reagents on the machines they will be used on (*see* **Note 10**).
- Order all that will be needed up front as long as the storage requirements do not exceed the estimated timeline for the screening process. If possible order the same lot of each reagent otherwise pool, aliquot, and store the reagent before running the test set and screen(s). Lot control is essential (*see* **Note 11**).
- Recheck the timing of the entire screen using the estimated times for each assay step in a screen: Often certain steps in a screen can create a bottle neck that creates longer incubation times for each subsequent plate. This needs to be looked out for and preempted (*see* **Note 12**).
- Perform a small automated run using a validation set. A validation set includes duplicate test plates which have random cDNAs, siRNAs, or miRNAs as well as the positive and negative controls spotted in triplicate per plate. Any subsequent changes in reagents, protocol, or technique should be followed by another test set before proceeding to a screen (*see* **Notes 13** and **15**).

3.4 Execute Screen

It is advised that each plate is run in duplicate and if the screen is to be run in multiple modes (plus and minus agonist for example) that for each plate both modes be run during that batch. For large screen sets, it is good practice to include control plates containing the same miRNA or siRNA reagent set in the beginning, the middle, and the end of the queue to compare during analysis for analysis of the screen quality at different time points in the screening queue.

- Prior to starting a screen protocol prepare all automation equipment by cleaning, calibrating, and testing to make sure the dispense amounts are precise and accurate.
- Prepare the library plates by spotting in the library and controls unless these have been spotted at an earlier date and the plates have been sealed and frozen. If the library was spotted and frozen, prepare the plates by thawing to room temperature with seal or cover intact; spraying with 70% ethanol; centrifuge at ~200 × *g* for 5 min (*see* **Note 14**).
- Spot controls into empty wells if necessary and cover each plate with a sterile lid.
- Execute the screening protocol optimized before without any deviations and note down any issues, failures or problems observed during the execution.

3.5 Data Analysis

- The first step in the data analysis is the assembly of the data. Typically, data are tracked by plate using automated identifiers such as barcodes.
- Check the data integrity for consistency of the time stamps. Typically each plate should have been acquired according to an automation schedule. Any deviation from the schedule points at potential problem.
- Visualize all data using an appropriate program. Heat maps—even in simple programs such as Excel are useful to visualize the single plate level. Detect any row or column drift by visualizing plate format heat maps of the data or histograms of row or column data. Data drift between plates can be detected by visualizing plate format heat maps, scatter plots, or box plots of plate data. Control wells can be particularly useful in identifying anomalies across plates (*see* **Note 15**).
- If drift is being detected, decide upon the reason and severity of the data drift within a plate the data might be rescued by normalizing the data to control wells or median data within the affected rows or columns (well data/row or column median). Otherwise the data should be discarded before proceeding to plate data normalization in order to prevent compromising the analysis of unaffected wells (*see* **Note 16**) or if severe drift is being detected, discard the plate's data to avoid compromising the analysis of unaffected plates.
- Normalize all plate data by dividing the individual well data by the median plate data. Alternately, each plate data can be normalized to a negative control. Due to the difficulty of finding a proper negative control for, for example, siRNA experiments and possible drift within the rows or columns containing the control wells, it is recommended to use the plate median rather

than the controls for plate data normalization rather than controls.

- Plot a scatterplot of the data with one replicate on the x axis and the other on the y axis. Calculate the linear correlation coefficient or Pearson product correlation coefficient to measure the strength of linear relation between the replicate data [2]. This number will reflect the noisiness of the screening data with a strong positive linear correlation indicating a higher confidence in the reproducibility of the replicate data.
- Rank each library members effect on the system, i.e., each cDNA's or siRNA's effect. One method to rank each reagent's effect is to calculate a signal to noise ratio. The raw data is normalized by dividing by the median value of the plate (excluding the control data). These normalized data are log transformed to yield symmetric ranges between fold changes above and below the median. Then the average fold activation (afa) of the normalized, log transformed values $x_1, x_2, \ldots, x_n$ is computed for the replicates n of each reagent (ie. a particular RNA construct or cDNA). The average fold activation is defined on the basis of a geometric mean of the replicate values.

$$\text{afa} = \begin{cases} \exp(\mu) & \text{if } \mu \geq 0 \\ \dfrac{-1}{\exp(\mu)} & \text{if } \mu < 0 \end{cases}, \text{with } \mu = \frac{\sum_{i=1}^{n} \log(x_i)}{n}$$

- Next the afa is penalized by dividing it by the adjusted standard deviation of the fold activation or median fold activation (mfa). The ratio afa/mfa is referred to as penalized fold activation (pafa). The mfa in effect penalizes the value for fold activation if the standard deviation between the replicates is high; increasing the likelihood of identifying reproducible hits.

$$\text{pafa} = \frac{\text{afa}}{\text{mfa}}, \text{with mfa} = \exp\left(\frac{\sqrt{\sum_{i=1}^{n} (\log(x_i) - \mu)^2}}{n - 1}\right)$$

- Visualized pafa values by scatter plot with the pafa values on the y axis and the reagent name on the x axis. If it is a repressor screen, then it is easier to visualize repressors which have a value between 1 and 0, by plotting the log value of the pafa. Alternately, each value below 1 can be converted to its negative inverse and hits be extracted (*see* **Note 17**).

3.6 Reconfirmation and Filter Screens

General considerations: A reconfirmation screen independently reproduces the experiment under the same conditions as the original screen using as many of the reagents (ie. siRNAs, miRNAs, or cDNAs) that have passed the statistical threshold for selection as experimentally feasible. Filter screens include the same reagents as the reconfirmation screen (ie. siRNAs, miRNAs, or cDNAs) but utilize additional experimental assays in order to further characterize the hits by determining whether they are specific to the biological process under interrogation. For example if the original screen had a luciferase read-out, the filter screens may include a cell titer read-out, a high-content imaging analysis, or perhaps a different cell type or reporter. It is important to include an assay here with a readout that is orthogonal to the readout used in the primary screen. The general workflow for follow up screening is given below:

- Rank data for individual reagents and prioritize the hits for follow-up via reconfirmation and filter screens (*see* **Note 18**).
- If the initial screen included two or more reagents per well in order to reduce the number of wells to screen, then in the reconfirmation screen, these reagents should now be tested individually in order to determine which was responsible for the phenotype.
- Design a plate layout with all of the reagents to be tested. Because reconfirmation assays contain a high number of initial outliers, data analysis can be skewed by these when normalizing the plate data by the mean. Therefore either a high number of random reagents must be included in these assay plates or the data should be normalized to a negative control within each plate rather than to the plate's mean data.
- Execute follow up screen.
- Collect the results and perform exit meeting with team to communicate the results and discuss further follow up.

4 Notes

1. Exclude bacterial or fungal contaminants: Bacterial or fungal contamination is easy to detect by eye or under a light microscope when antibiotic and antimitotic reagents are not being used. Antibiotic and antimitotic reagents are not normally necessary for routine cell culturing and should only be used under high risk situation. This is because they may alter the biology of the cells, encourage cryptic contamination, and conceal poor aseptic technique [3].

2. Exclude mycoplasma contamination: The main sources of mycoplasma contamination in a cell culture laboratory are animal-derived media products, laboratory personnel, and cross contamination of other contaminated cell lines. The lack of a cell wall in mycoplasmas makes them invisible to the naked eye [4]. The side-effects of mycoplasma contamination on cell cultures are (1) inhibition of proliferation, (2) enhanced cell death, (3) fragmentation of DNA and (4) morphological features of apoptosis [5]. There are several methods for detecting mycoplasmas including the microbiological culture method, DNA staining by fluorochromes, PCR, fluorescence in situ hybridization (FISH), and assays based on the detection of adenosine triphosphate (ATP) generation by fluorescence microscopy and luminometer [4].
3. Verify the cell line: It is important to validate the cells which will be used against a previously authenticated stock either in a cell bank or the laboratory of origin and also verify that it is free from contamination [6]. Cell line validation should be done prior to screening. Many publicly available protocols and commercially available kits and services are available for cell line validation. Consult regulatory requirements or publication requirements in order to determine the correct methods to employ for verification.
4. It is typically necessary to have many aliquots of the cell line at the same low passage number which can be seeded in advance of a screen so that the cells are at the appropriate passage when the screen will be performed. It is necessary to consistently record variables and observations such as morphology, so that the optimal conditions can be recognized. In order to minimize the risk of jeopardizing large data sets, ensure that passaging protocol, passaging numbers, and media age and lot numbers are constant between test sets and actual screens. In addition, selection agents used to maintain stable plasmids and any other unnecessary reagents should be removed from cell culture at least 1 week prior to using the cells for a test set or screen in order eliminate complications to the data due to these reagents' effect on cell biology.
5. In functional genomics, the screening libraries are typically prearrayed into plates with for example one siRNA and one cDNA per well. Fastidious tracking of plate maps is necessary. The screen will use a reverse transfection approach: With reverse transfection, transfection mix containing a transfection reagent and serum-free media is dispensed into the well containing cDNA, siRNA, or miRNA and incubated while complexes between the transfection reagent and the cDNA, siRNA, or miRNA form, then the cells in media are added.

Transfection reagents and protocols with miRNAs are generally the same as transfections of siRNAs. cDNAs usually require different transfection reagents than siRNAs and miRNAs though there are some commercially available transfection reagents that claim to work with cDNA, siRNA, and miRNA. To determine whether a particular transfection reagent will work with your assay and subsequently optimize transfection, use for example a GFP-expressing plasmid or a siRNA that represses for example luciferase of GFP and also check with positive and a negative control cDNA, siRNA, or miRNAs for phenotypes. Within the same assay plate, test all available transfection reagents at two to three different concentrations within the range recommended by the reagent's vendor in quadruplicate. This initial optimization will help determine the best reagent and concentration. This should be followed by additional optimization assays to refine the protocol. Choose the transfection reagent and concentration with that yields the greatest average difference in activity between the positive and negative control and with the least amount of cytotoxicity or other measurable nonspecific effects in the negative control. Some cell types cannot be efficiently transfected using any currently available transfection reagent. In this case a viral delivery system should be considered.

6. The biological system under analysis as well as reagent requirements and limitations are considered when devising the protocol timeline. Generally, the effects of miRNAs and siRNAs can be observed by 48–72 h. The actual time for individual miRNAs and siRNAs varies and is dependent on the stability of the protein whose mRNA is being regulated. Unfortunately, when screening a large library it is usually not feasible to tailor the screen length to individual protein stabilities. Additionally, some screens may require more time between miRNA or siRNA transfection and endpoint due to the time requirements for certain phenotypes to appear. In contrast, the effect of cDNA expression is more immediate and cDNA can generally be observed between 24 and 48 h though the endpoint time can also vary due to the time requirements for certain phenotypes to appear. When a screen is probing the effect of knocking down (miRNA or siRNA) or overexpressing a gene (cDNA) on another reagent's, such as a compound's, ability to induce a phenotype, then it is prudent to give the miRNA, siRNA, or cDNA time to work before adding the compound.

7. It is difficult to find a true negative control for RNAi. The complementarity of as few as 7 nucleotides of a siRNA or microRNA can already determine the knock down of relevant transcripts [7]. A close comparison of untreated to control siRNA treated cell cultures will help to identify the right

control molecules, which show a controlled or even show no impact on the assay system and will be appropriate for the normalization of the experimental observations [8]. Positive controls could include any miRNA, siRNA, or cDNA already known to result in the desired phenotype. Other positive controls could include miRNA, siRNA, or cDNA known to induce cell death or any other measurable phenotype in the cell type being interrogated or fluorescently tagged miRNA, siRNA, or cDNA. These positive controls would be useful for assurance that the transfection was successful.

8. A negative control for a cDNA screen could include the parent plasmid(s) included in the cDNA library containing either no cDNA or a cDNA that does not affect the assay.
9. After determining the preferred cell density, it is important to take steps to minimize variance from this density between wells, plates, and batches. These steps include consistent cell passaging, harvesting, counting, and dispensing techniques. Steps to minimize or eliminate cell clumping should also be employed. The protocol timeline should also be considered when determining the optimal cell density. The time required for the phenotype to appear must be carefully balanced with the cells' growth rate. The cell density needs to be optimized to the assay.
10. Many reagents such as transfection reagents, cell preparations, and luminescence reagents have decreased performance if not used within a certain timeframe. These reagents will need to be prepared in batches when larger screens are being performed. To test whether this will be necessary, test the reagents' performance after being left out for different time points under screening conditions, i.e., sitting on the machines.
11. Take into consideration that some reagents will have to be replenished during the screen as they do decay over time while sitting on the machine. Also some machines have dead volumes which must be accounted for.
12. For example if dispensing a reagent such as formaldehyde takes 1 min to be dispensed, then after a 15 min incubation needs to be washed off with a 3 min wash step, it will be necessary to incorporate a 2 min incubation linked prior to each formaldehyde dispense in order to prevent a bottleneck which leads to subsequent plates being exposed to increasing incubation times with formaldehyde.
13. The test set should be performed using the same conditions that will be used for the screen. The test set is useful for showing how the controls perform in relation to a larger random data set and for showing whether the phenotype of individual random cDNAs, siRNAs, or miRNAs and their triplicate

data have large or small variance. The test set is also a "dress rehearsal" for the screen, and can illuminate possible problems with communication between collaborators when multiple people are involved in preparing reagents, or possible problems with the automation or protocol that will be used for the screen.

14. When working with siRNA or miRNA, wipe with RNAseZap or similar reagent and remove seal.
15. Although all attempts should be made to reduce data drift within individual plates or variation between plates, it is possible and even likely that signal drift problems will occur, especially when working with a large batch of plates. Data drift between plates is generally due to the activity of reagents decreasing or increasing during the time between dispense of the first and last plate. This drift can be identified by comparing control well values across all plates within the assay. As discussed previously, this drift can be minimized by knowing the behavior of each reagent over time and preparing the reagents in batches so that each plate receives reagent that is optimally active. In addition to time, the activity of a reagent can change at different temperatures. Unless a reagent is kept constant within a water bath during dispenses, it is prudent to bring each reagent to room temperature before beginning a dispense step. A bottleneck at a particular step in the screening protocol which contributes to variations in incubation or storage times can also contribute to data drift between plates.
16. Data drift within plates (row or column drift) can occur for several different reasons. Most often it occurs due to problems that arise with the dispensing or aspirating equipment being used. Despite calibration and testing prior to protocol initiation, dispenser and/or aspirator tips can become partially or fully clogged or have decreased function during the screen. Aspirator and dispenser tips can become clogged due to reagent build-up. Thoroughly cleaning equipment before and after using the equipment can reduce build-up of reagents and prevent clogs. Dispenser clogs can also be reduced by filtering reagents and cells to remove clumps and large particulates before dispensing. Aspirator tips can be clogged by mold that may come up in some wells or plates during incubation steps. Maintaining a sterile working environment by HEPA filtering, controlled airflow, routine monitoring, and sterilization are key to preventing mold. In addition to reagent aspirators and dispensers, imagers or detectors used to collect the data at the end of an assay can contribute to data drift within plates. This could be due to improper calibration, the plates not resting properly in their holder, or misalignment of the plate. Other row or

column drift can arise due to any number of other mechanical issues that depend upon the specific equipment employed for a particular screen.

17. Among others, an alternate statistical method called strictly standardized mean difference (SSMD) has been proposed to measure the effect represented by the magnitude of difference between a reagent and a negative reference group [9]. New analytical methods should continually be evaluated as they are published in order to maximize the value of the data generated by screening efforts.
18. Depending on resources, additional coverage may be added to the reconfirmation screen in order to increase information that can aid in prioritizing hits for further investigation. For example, if the original screen was an siRNA screen then additional constructs to the hit targets may be included in order to increase confidence in the results. If the original screen did not include enough reagents in order to cover the genome, then the hit list can be functionally analyzed using bioinformatics resources which probe the hit list for pathway or gene family enrichment which may reflect underlying molecular processes. Reagents related to these disease pathways or gene families that were not included in the original library can be added to the reconfirmation screen in order to find additional hits that would have been missed or increase confidence in the original results. Many commercial and publicly available analytical tools for determining gene enrichment are available including, but not limited to, QIAGEN's Ingenuity® Pathway Analysis (IPA®, QIAGEN Redwood City, www.qiagen.com/ingenuity), The Database for Annotation, Visualization, and Integrated Discovery (DAVID) [10].

References

1. König R et al (2007) A probability-based approach for the analysis of large-scale RNAi screens. Nat Method 4(10):847–849
2. Sullivan M III (2006) Statistics: informed decisions using data, 2nd edn. Pearson Prentice Hall, Upper Saddle River
3. Freshney R (2002) Cell line provenance. Cytotechnology 39(2):55–67. https://doi.org/10.1023/A:1022949730029
4. Nikfarjam L, Farzaneh P (2012) Prevention and detection of mycoplasma contamination in cell culture. Cell J 13(4):203–212
5. Sokolova IA, Vaughan AT, Khodarev NN (1998) Mycoplasma infection can sensitize host cells to apoptosis through contribution of apoptotic-like endonuclease(s). Immunol Cell Biol 76(6):526–534
6. Geraghty RJ, Capes-Davis A, Davis JM, Downward J, Freshney RI, Knezevic I, Lovell-Badge R, Masters JR, Meredith J, Stacey GN, Thraves P, Vias M (2014) Guidelines for the use of cell lines in biomedical research. Cancer Research UK. Br J Cancer 111(6):1021–1046. https://doi.org/10.1038/bjc.2014.166
7. Lin X, Ruan X, Anderson MG, McDowell JA, Kroeger PE, Fesik SW, Shen Y (2005) siRNA-mediated off-target gene silencing triggered by a 7 nt complementation. Nucleic Acids Res 33:4527–4535
8. Baum P, Fundel-Clemens K, Kreuz S, Kontermann RE, Weith A, Mennerich D, Rippmann JF (2010) Off-target analysis of control siRNA molecules reveals important differences in the cytokine profile and inflammation response of

human fibroblasts. Oligonucleotides 20 (1):17–26. https://doi.org/10.1089/oli.2009.0213

9. Zhang XD, Ferrer M, Espeseth AS, Marine SD, Stec EM, Crackower MA, Holder DJ, Heyse JF, Strulovici B (2007) The use of strictly standardized mean difference for hit selection in primary RNA interference high-throughput screening experiments. J Biomol Screen 12 (4):497–509

10. Huang da W, Sherman BT, Lempicki RA (2009) Systematic and integrative analysis of large gene lists using DAVID bioinformatics resources. Nat Protoc 4(1):44–57. https://doi.org/10.1038/nprot.2008.211

Chapter 8

Reporter Gene Assays Using Transfectable Functional Genomics Libraries

Genevieve Welch, Robert Damoiseaux, and Loren Miraglia

Abstract

Transfectable functional genomics libraries are traditionally the workhorses of functional genomics screening using reporter gene assays. These libraries offer insight into fundamental cellular processes governing health and disease and can be utilized in an arrayed fashion which makes them uniquely suited to deconvolute complicated disease phenotypes and dissect biological networks that would otherwise be inaccessible. Here we give an overview of the principles for the generation, screening and data analysis of such arrayed libraries. Specifically we cover the differences between the various transfectable reagents, library selection and handling, and data analysis to offer a comprehensive understanding of these important technologies and how to apply them.

Key words Functional genomics, siRNA, miRNA, cDNA, HTS, High-throughput screening, Transfection, Libraries, Data analysis, Automation

1 Introduction

In the beginning of functional genomics, only cDNA libraries were available to dissect biology using reporter gene assays. This changed drastically with the advent of interfering/silencing siRNA and eventually micro RNA mimics. All three of them offer powerful and useful reagents that enable the researcher to gain access to different and unique aspects of biology that is not available through one of the other reagents alone.

Specifically, silencing reagents and screens offer different sets of hits from for example cDNA screens, much like the loss of function and gain of function or positive and negative modulation in the cell have different effects that can not be substituted for the other. At this point, the reagents are widely commercially available but the know-how and technology of their use is unfortunately still less than well documented. The same goes for the curation of the reagents, handling as well as their overall characterization. Specific issues such as drifts in the annotation of the genome that leaves

Robert Damoiseaux and Samuel Hasson (eds.), *Reporter Gene Assays: Methods and Protocols*, Methods in Molecular Biology, vol. 1755, https://doi.org/10.1007/978-1-4939-7724-6_8, © Springer Science+Business Media, LLC, part of Springer Nature 2018

come clones without recognized targets, issues with fidelity of the libraries during replication in for example cDNA libraries or solubility issues of siRNA libraries can make the use of these resources challenging. Here we will describe our approach to the use of these libraries and also provide guidance on the curation of the libraries and proper quality controls to be put into place to ensure success and reproducibility of the results.

Our approach differs from the standard one-compound one-replicate screening methodologies as generally functional genomics screens are run in at least duplicate. Moreover, the methodologies employed here put a premium on a comprehensive execution of the assay development workflow lined out in the previous chapter (*see* Chapter 7) as the assay can not be fixed or improved during the actual screening process. We will start with a discussion of each reagent type and how to select the proper reagent, then discuss the copying and handling of the library and finally show an example of a miRNA screening protocol. The actual screening protocol is with minor modifications such as transfection reagent type as well as optimization as outlined in the previous chapter transferrable to the other reagent types. The first step in the use of functional genomics screening is to select the library/reagent type after a target has been greenlighted for screening.

2 Materials and Methods

2.1 Protocols

This section starts with protocols for the selection and handling of libraries and then a practical sample protocol for the screening for miRNA/cDNA/miRNA is shown to exemplify how these libraries are applied in specific.

2.2 Selecting a cDNA Library

A cDNA (complementary DNA) library is composed of plasmids containing cDNA whose expression is controlled by a constitutive or inducible promoter. cDNA is produced from fully transcribed mRNA (mature RNA) which lack enhancers, introns, and other regulatory elements found in a genomic DNA library. Also because alternate splicing events or mutations can occur to create an mRNA, it is necessary that they are fully sequenced. If the resources required to sequence the entire library are not available, then it is acceptable to simply sequence the cDNA hits. It is prudent to sequence from the same well of a copy plate of the library that was screened. This can be achieved by transforming a small amount of the cDNA into bacteria, then expanding, miniprepping, and sequencing the plasmid. The reason for doing this is that if the working library plates were misannotated or the if plate was inverted when spotting the working plates, then the actual hit will be determined or confirmed before valuable resources are expended for follow-up on the wrong cDNA. Misannotations are actually quite frequent in cDNA screening. It is relatively difficult to copy a cDNA library with 100% fidelity because even miniscule contaminations can lead to cross-contamination of wells within the same plate. Below we will give our protocol for replication of cDNA libraries, but even using the best precautions, it is imperative, that cDNA libraries be quality controlled, i.e., sequence verified. Selecting and handling a cDNA library is outlined in a step by step manner below:

1. Chose the correct library background: the parent plasmid containing the cDNAs is an important decision. Also, it puts restrictions on the usability of the cDNA library. One restriction is that the promotor of the library has to be transcribed in the cellular target background. A second restriction can come in the form of the size of the plasmid. Large plasmids transfect worse making difficult to transfect cell lines harder to work within the context of a large library plasmid.
2. Determine the necessary tools for the screen. Frequently library plasmid contain a multitude of tools such as GFP to mark transfected cells or a selectable marker to isolate pure populations.
3. Consider possible plasmid toxicity. This can be a function of the plasmid background and should be assessed experimentally in the cell background in question. Normally, this is also a function of dose applied per well. With larger plasmid size, the amount of DNA needed to achieve the same effect goes up. Smaller plasmids make more sense for most applications.
4. Check the coverage: The clone size of a given plasmid library is only a rough estimate of the actual coverage. Analyze the number of targets and ensure that the biological pathways in question are covered. Also, cross-check the history of the annotation if bioinformatics resources are available.
5. In cDNA libraries the species frequently plays a smaller role than in for example siRNA libraries where nucleotide sequences targeting mouse genes as for example may not work in human cells and vice versa. However, this is not always true. Check to ensure that your most important proteins are present in the correct species.
6. Miniprep a subset of the clones and sequence to ensure library integrity before committing to a screen.

2.3 Selecting an siRNA Library

There are many commercially available siRNA libraries that differ in the algorithms used for predicting siRNA sequences and the length and modifications of their construct sequences. All three of these parameters can impact off target liabilities [1, 2]. It is often difficult to predict which library would be the best purchase given all the available options. Features of the library such as cost, licensing agreements, vendor validation of constructs' on-target affects are factors. Other factors include low cytotoxicity, and increased knock-down efficiency that is frequently achieved through modification of the nucleotides. If possible, select a small set of genes and compare that set targeting the same genes from different vendors before making a large purchase. We will discuss below the overall handling and QC process for an siRNA library:

1. When receiving a library, check that all plate seals are still intact and that all plates are intact as well.
2. Spin down the plates in an RNAse-free centrifuge (*see* Subheading 2.9 below for details on plate handling) and deseal in a biosafety cabinet.
3. Add RNAse free siRNA suspension buffer.
4. Reseal plates with RNAse free plate seals.
5. Vortex each plate, spin down, and freeze plates at −80 °C overnight. Repeat this process for a total of at least three times.
6. Determine the RNA concentration for a subset of wells. They should be within 30% of specification. Consider that different nucleotide sequences have different absorbance values. If need be, calculate the Lambert–Beer coefficient for each construct separately.
7. If <10% of the wells are out of specification, then the library is validated for screening.

2.4 Selecting and Handling a miRNA Mimic or miRNA Inhibitor Library

In miRNA screens, individual miRNA mimics and inhibitors are transfected into cells to induce or inhibit a specific phenotype. Currently hundreds of miRNAs have been characterized for each common experimental species. There are several commercially available collections which incorporate different chemical modification strategies for the nucleotides. The sequences of and the chemical modifications to the miRNA mimics or inhibitors often differ between each library and will have varying degrees of effectiveness for mimicking or quenching endogenous miRNAs. It is prudent to carefully research and test a library's functionality before making a large purchase which should rely on the resulting data.

miRNA mimic collections are used to systematically introduce and/or overexpress miRNAs into a cell of interest in order to find miRNAs which illicit a desired phenotype which is a very powerful technology useful for multiple applications. The function is similar to the siRNA technology, but the data analysis can be more complex. In addition, phenotypes can vary depending on the cell type of interest and the experimental conditions and due to this it is important to carefully plan the cell type and the experimental conditions including cell density. Some drawbacks to using mimics are the possibility of overloading the limited availability of miRNA-Dicer or RISC complex structures [3] thus dampening the effects of the miRNAs already present in the cell. Finally, because a miRNA is being artificially introduced, this does not shed light upon the function of the miRNAs normally present in the cell under those conditions.

Conversely, miRNA-inhibiting strategies such as antagomiRs, locked nucleic acids, or antisense oligonucleotides are used to suppress the function of miRNAs that are present under specific experimental conditions and increase the expression of their target gene(s). This method for finding miRNAs which are important to a biological pathway can be more useful for finding miRNAs that are actually present in the probed biological system. As with miRNA mimic screens this phenotype can vary depending on the cell type of interest and the experimental conditions. Therefore it is important to carefully plan the cell type and the experimental conditions including cell density.

If a double-stranded hairpin miRNA library is used, it is possible for two different mature miRNA sequences to be excised from opposite arms of the same hairpin precursor. Further analysis will be needed to determine whether the miRNA from the 5′ or 3′ arm of the hairpin is responsible for the phenotype. The handling of these libraries is for all intended purposes identical to the handling of siRNA libraries.

2.5 Library Spotting or Stamping

Depending on the library acquired, it may be necessary to make a working copy and dispense small aliquots into the destination plates at working concentration. This is often referred to as stamping or spotting. These plates can then be frozen at the recommended temperature for later use. Spotting multiple copies of a library reduces the amount of time it takes to run subsequent screens by eliminating the need to do this each time. It also preserves the library by reducing the need to repeatedly freeze, thaw, and expose the reagents and signficantly reduces costs if disposable tips are used for the initial spotting.

Generally destination plates will be a 96-well, 384-well, or 1536-well plate. These plates could be white solid for luminescence assays or black with clear bottom for imaging. Other options such as coating depend on the cell requirements and the instruments to be used. Ideally, the plates should be spotted using an automated dispenser which places the drop directly at the bottom of the well to reduce variability and mistakes.

Barcodes becomes very important to enable tracking of plates and matching it up with the inventory plate maps of the reagents (*see* **Note 1**). It may be useful to leave multiple wells empty, as these locations could be used for adding control miRNAs or siRNAs. It is also recommended to do thorough quality assurance before attempting to spot the library to test the accuracy and precision of the dispenser. This can be confirmed by spotting a luminescent dye such as FITC, and determining the variability between wells. Additionally, the durability of all materials and reagents during the freezing and thawing process should be confirmed experimentally before spotting valuable reagents otherwise the plates may break or the seals may come undone upon thawing exposing the wells to contamination or condensation. This is especially important when considering the employment of automated heat sealers. If the temperature is not high enough on the sealer, the plate seal may warp off over time, but if the sealing temperature is too high, it may damage the spotted reagent during sealing. Often, this problem takes a long time to become evident as it will be only visible after thawing a copy. Consulting with well-established screening laboratories can help with selecting vetted reagents.

Another important factor to consider when spotting a library is positional bias that occurs within a plate during the screening process. Although running replicates of each plate helps minimize the effect of data abnormalities, when this abnormality is consistent to a specific plate position across plates the data may incorrectly appear to have low variability. In the context of HTS, randomly assigning reagents to different well locations across replicated plates would remove confounding of biological activity with well location and associated extraneous variables such as evaporation, air humidity, UV exposure, incubator stack position, and liquid handling [4]. As a result, the measure of variability would be more accurate and any positional biases would be spread across reagents. To do this, the researcher would need to create two parent versions of the library before spotting, using random well assignments generated via random number algorithms to rearray the wells for the second version. If the resources for such a prudent step is not available, then it is even more important for positional biases to be monitored and avoided. Sentinel plates can be placed throughout the screen which contain the same reagent in each well as a means of determining whether positional bias has occurred and roughly when in the screen the problem arose.

Moreover, empty wells should be left in each destination plate so that assay specific controls may be added as necessary. Ideally these would be located in different locations across replicate plates for reasons described above.

2.6 Spotting Protocol for Transfectable Libraries

1. Select the plate types and number of copies to be stamped.
2. Barcode the plates with barcodes containing date, ID, and a reference to the plate map in the inventory.
3. Validate the precision and repeatability of the spotting (*see* **Notes 2** and **3**).
4. Thaw out only the number of plates and reagents that can be processed in a single run.

5. Rearray wells to a working copy if possible leaving control wells empty. At this point the spotting process could be interrupted if the library plates are frozen down.
6. Spot the copies. This should be performed using an automated liquid handling system that scans barcodes and generates a trace for each pipetting transfer. This will aid with the recovery in case any irregularities are ever found.
7. Seal the copies with a plate seal that has been validated.
8. Rack plates and freeze at −80 °C. Racking plates is essential for the purpose of inventory control but also usability. For example, if each rack contains one copy, accessing a library will become much easier.

2.7 Screening Transfectable Functional Genomics Libraries

Transfectable functional genomics reagents have a similar protocol and a similar optimization path but it is especially important to consider the concentration of miRNA in the transfection. Determine the highest concentration without off target effects when comparing negative control siRNA transfected cells to mock transfected cells. Generally this is done by comparing the data between cells transfected with scrambled siRNAs and mock-transfected cells (transfection protocol minus siRNAs). Concentrations that have the greatest affect using positive controls but the least amount of variability between scramble siRNA transfected and mock transfected cells should be selected. The nonspecific effects on gene expression are dependent upon siRNA concentration in a gene-specific manner [5]. Obviously the optimal concentration will vary from screen to screen based upon the transfection efficiency and number of cells among other variables. If multiple sets will be spotted without foreknowledge about the assays for which they will be used, it is recommended to optimize concentration in a cell line with high transfection efficiency.

2.8 Practical Screen Sample Protocols for miRNA/cDNA/miRNA

The reagents needed for all screens are rather similar with the exception of the transfection reagent flavor and possibly the controls if one wants to use for example an siRNA control for a miRNA screen and a cDNA for a cDNA screen. The actual reagents named and the machines used can differ from laboratory to laboratory but the combinations below allow for a solid performance. Substitutions are possible where indicated.

Reagents

T175 flask.

Hek 293T cells.

miRNA mimic library.

White Solid 384-well plates.

Plasmid Transfection Reagent such as FuGENE® 6 (Promega).

Opti-MEM® I Reduced Serum Medium, no phenol red (Gibco®).

TNK1 cDNA plasmid (endotoxin-free).

STAT4(2)-luc plasmid (endotoxin-free).

IFNgamma.

FBS.

DMEM, high glucose, HEPES, no phenol red (Gibco®) cat # 21063-029.

DPBS.

TrypLE™ Express (Invitrogen Life Science Technologies).

Hemocytometer.

Trypan Blue.

Cell death siRNA.

Scramble siRNA.

Lipofectamine® RNAiMAX Transfection Reagent (Invitrogen Life Science Technologies) or similar product.

Beetle Luciferin, Potassium Salt (Promega) cat # E1605.

CellTiter-Glo® Luminescent Cell Viability Assay (Promega) cat # G7572.

EnVision® Multilabel Reader (PerkinElmer) Product number: 2104-0010.

GNF Systems Dispenser.

Bravo Automated Liquid Handling Platform (Agilent).

15 mL conical tubes.

50 mL conical tubes.

RNaseZap® (Invitrogen Life Science Technologies).

2.9 miRNA Screening Protocol

The objective of this screen was to find miRNA mimics that agonize or antagonize the STAT4 promoter in Hek293T cells. In order to find miRNA mimics that agonize the reporter (agonist assay) (Assay #1) we transfected miRNAs into Hek293T cells that were transiently transfected with a luciferase promoter driven by STAT4 promoter in order to find miRNAs that increased the signal. In order to find miRNA mimics that antagonize the reporter (antagonist assay) we transfected miRNAs into Hek293T cells that were transiently transfected with a luciferase promoter driven by STAT4 promoter and agonized by either It has been demonstrated that in diverse animal cell lines, the global efficiency of miRNA biogenesis is intimately linked to cell density. As diverse mammalian and Drosophila cell lines are grown to increasing density, miRNA biogenesis is globally activated, leading to elevated mature miRNA levels and stronger repression of target constructs. This broad increase in miRNA abundance is associated with enhanced processing of miRNAs by Drosha and more efficient formation of RNA-induced silencing complexes [6]. These findings reveal that careful monitoring of cellular confluency is critical for accurate analysis of miRNA expression and function in widely used cell culture systems. One can use for example 0.3 ng/mL IFNgamma (*see* assay #2) or TNK1cDNA (*see* assay #3) in order to find miRNAs that decreased the signal. Reporter signal luminescence is typically collected at three time-points, then cell viability will be assessed to filter out signal effects not specific to reporter signal.

Reagents

T175 flask.

Hek 293T cells.

miRNA mimic library.

Greiner White Solid 384-well plates.

FuGENE® 6 Transfection Reagent (Promega).

Opti-MEM® I Reduced Serum Medium, no phenol red (Gibco®).

TNK1 cDNA plasmid (endotoxin-free).

STAT4(2)-luc plasmid (endotoxin-free).

IFNgamma.

FBS.

DMEM, high glucose, HEPES, no phenol red (Gibco®) cat # 21063-029.

DPBS.

TrypLE™ Express (Invitrogen Life Science Technologies).

Hemocytometer.

Trypan Blue.

Cell death siRNA.

Scramble siRNA.

Lipofectamine® RNAiMAX Transfection Reagent (Invitrogen Life Science Technologies).

Beetle Luciferin, Potassium Salt (Promega) cat # E1605.

CellTiter-Glo® Luminescent Cell Viability Assay (Promega) cat # G7572.

EnVision® Multilabel Reader (PerkinElmer) Product number: 2104-0010.

GNF Systems WDII.

Bravo Automated Liquid Handling Platform (Agilent).

15 mL conical tubes.

50 mL conical tubes.

RNaseZap® (Invitrogen Life Science Technologies).

2.10 Practical Screen Sample Protocols for cDNA Using Transient Transfection of 293T Cells with cDNA Reporter and cDNA Agonist

1. Seed low passage number Hek293T cells (~18 million) in 10% FBS DMEM to a T175 flask so that the cell density will be approximately 80% confluent the next day. Prepare 1 flask for each assay.
2. Incubate at 37 °C, 5%CO_2, 95% humidity for 24 h.
3. Prepare transfection cocktail in sterile 14 mL tubes by aliquoting 2.34 mL OptiMEM then adding 15 μg of STAT4-luc reporter plasmid for assays #1–3 and 7.5 μg of TNK1 cDNA for assay #3. Add 3 μL of FuGENE® 6 Transfection Reagent for every microgram of DNA and mix.
4. Incubate transfection cocktail at room temperature for 20–30 min.
5. Remove the media from the flasks containing 80% confluent Hek293T cells and replace with 16.6 mL of 2% FBS DMEM no indicator.
6. Add transfection reagent drop wise to the flasks and mix by swirling.
7. Incubate at 37 °C, 5%CO_2, 95% humidity for 18 h.

2.11 Transient Transfection with miRNAs and Data Collection

1. Preparation of miRNA library plates which were previously spotted with 1 μL of 0.4 μM miRNA per well into white solid 384-well plates, foil-sealed, and stored at −80 °C.
   (a) Thaw two copies per assay of the miRNA library plates to room temperature.
   (b) Spray the plates with 70% ethanol solution.
   (c) Centrifuge the plates at ~1120 × g for 5 min.
   (d) Wipe the plates with RNaseZap®.
   (e) Remove the foil seals and spot 1 μL of 1 μM control cell death and scramble siRNAs into the empty wells reserved for controls by using a liquid handler such as Agilent Bravo Automated Liquid Handling Platform.
   (f) Place lids on the plates.
2. Cell Preparation.
   (a) Aspirate the media from the T175 flasks containing cells transfected with cDNAs.
   (b) Rinse each flask with 5 mL DPBS then aspirate to remove.
   (c) Add 3 mL of TrypLE™ Express, swirl to coat cells, and incubate at 37 °C until the cells have detached from the flasks.
   (d) Add 15 mL 4% FBS DMEM, high glucose, HEPES, no phenol red to each flask and transfer the cells for each assay to a separate 50 mL conical tube.

(e) Pellet cells in a centrifuge at ~200 × *g* for 5 min. Meanwhile prepare and dispense the transfection cocktail.

(f) Remove media containing TrypLE and resuspend in 15 mL 4% FBS DMEM, high glucose, HEPES, no phenol red.

(g) Filter cells using a 20 μm filter to remove clumped cells.

(h) Remove a small aliquot of each cell sample, add Trypan Blue and count the viable cells using a hemocytometer.

(i) Dilute cells to 500,000 cells/mL in 4% FBS DMEM, high glucose, HEPES, no phenol red, 60 μM Beetle Luciferin (all assays), 0.6 ng/mL IFNgamma (assay #2 only).

(j) Using a stir bar keep a homogenous cell mixture until and during dispense.

3. Transfection.

(a) Using a liquid handler such as GNF Systems WDII, dispense 20 μL of transfection cocktail (0.045 μL Lipofectamine® RNAiMAX Transfection Reagent/20 μL Opti-MEM® I Reduced Serum Medium, no phenol red) to each well of the library plates.

(b) Incubate plates at room temperature for 20 min.

(c) Using a liquid handler such as GNF Systems WDII, dispense 20 μL of cell mixtures to two copies of each library plate per assay condition.

(d) Incubate at 37 °C, 5%CO_2, 95% humidity for 24 h.

4. 24 h time point reporter luminescence.

(a) Measure reporter luminescence using a 384-well plate detector luminescence detector such as EnVision® Multilabel Reader.

(b) Incubate at 37 °C, 5%CO_2, 95% humidity for 24 h.

5. 48 h time point reporter luminescence.

(a) Measure reporter luminescence using a 384-well plate detector luminescence detector such as EnVision® Multilabel Reader.

(b) Incubate at 37 °C, 5%CO_2, 95% humidity for 24 h.

6. 72 h time point reporter luminescence.

(a) Measure reporter luminescence using a 384-well plate detector luminescence detector such as EnVision® Multilabel Reader.

(b) Incubate at 37 °C, 5%CO_2, 95% humidity for 24 h.

7. 72.5 h time point Cell Titer.

(a) Using a liquid handler such as GNF Systems WDII, dispense 30 μL of CellTiter-Glo® per well.

(b) Measure reporter luminescence using a 384-well plate detector luminescence detector such as EnVision® Multilabel Reader.

(c) Safely discard plates.

2.12 Additional Analysis for siRNA Libraries Containing Multiple Constructs per Gene Target

If the screen is an siRNA screen and multiple siRNA constructs per target gene are included in the siRNA library, then prioritizing the hits is complicated by the variation in activity levels between the different constructs to the same target gene and the possibility that any effect may be off target due to the construct targeting a different gene instead of or in addition to its intended target. If the preliminary reconfirmation list needs to be limited to a certain number of constructs that is less than the number of constructs that meet the effect criteria for being an outlier, then the targets must be prioritized in a way that reduces the number of false positives thus increasing the resources to include those that may be false negatives. Prioritizing targets could be done by ranking the individual siRNA constructs by their pafa and selecting the largest number of potent siRNAs that is practical for confirmation and validation assays. This hit ranking method prioritizes genes targeted by a single more active siRNA which may be due to off-target effects or experimental error and increases the likelihood of missing genes targeted by multiple less active siRNAs which are less likely to have occurred by chance and are thus more likely to be reconfirmed. One method for prioritizing targets for reconfirmation by enriching for targets that have effects using multiple siRNA constructs is redundant siRNA activity (RSA) analysis. In RSA all wells in an assay are initially ranked according to their signals. Then, the rank distribution of all siRNAs (wells) targeting the same gene is examined and a *P*-value is assigned. Thus, *P*-value indicates the statistical significance of all wells targeting a single gene being unusually distributed toward the top ranking slots, calculated based on an iterative hypergeometric distribution formula (Supplementary Methods online; http://carrier.gnf.org/publications/RSA) [7].

2.13 Reconfirmation and Filter Screens

After the data for individual reagents have been ranked, then the screener is ready to prioritize the hits for follow-up via reconfirmation and filter screens. A reconfirmation screen independently reproduces the experiment under the same conditions includes as the original screen using as many of the reagents (i.e., siRNAs, miRNAs, or cDNAs) that have passed the statistical threshold for selection as experimentally feasible. Filter screens include the same reagents as the reconfirmation screen (i.e., siRNAs, miRNAs, or cDNAs) but utilize additional experimental assays in order to further characterize the hits by determining whether they are specific to the biological process under interrogation. For example if the original screen had a luciferase readout, the filter screens may include a cell titer readout, a high-content imaging analysis, or perhaps a different cell type or reporter.

If the initial screen included two or more reagents per well in order to reduce the number of wells to screen, then in the reconfirmation screen, these reagents should be arrayed individually in order to determine which was responsible for the phenotype.

Depending on resources additional coverage may be added to the reconfirmation screen in order to increase information that can aid in prioritizing hits for further investigation. For example, if the original screen was an siRNA screen then additional constructs to the genes of interest might be included in order to increase confidence in the results. If the original screen did not include enough reagents in order to cover the genome, then the hit list can be functionally analyzed using bioinformatics resources which probe the hit list for pathway or gene family enrichment which may reflect underlying molecular processes. Reagents related to these disease pathways or gene families that were not included in the original library can be added to the reconfirmation screen in order to find additional hits that would have been missed or increase confidence in the original results. Many commercial and publically available analytical tools for determining gene enrichment are available including, but not limited to, QIAGEN's Ingenuity® Pathway Analysis (IPA®, QIAGEN Redwood City, www.qiagen.com/ingenuity), The Database for Annotation, Visualization, and Integrated Discovery (DAVID) [8].

Due to the fact that reconfirmation assays contain a high number of initial outliers, data analysis can be skewed by these when normalizing the plate data by the mean. Therefore either a high number of random reagents must be included in these assay plates or the data should be normalized to a negative control within each plate rather than to the plate's mean data.

3 Notes

1. It is recommended to barcode with an odd number on the west side of the plate and an even number on the east side of the plate. This allows for tracking the plate, aligning each plate to its plate map, and this gene annotation also aids in ensuring that the plate was not mistakenly inverted at any time during the spotting or screening process.
2. Validation of the volumes for spotting of libraries is different from liquid to liquid transfer. Even if a procedure has been validated to liquid to liquid transfer, dry touch off in a well will frequently produce different results and needs to be optimized.
3. Liquid handling equipment needs to be validated on a regular base. Pipetting using fluorescent tracers such as fluorescein will enable validation of very small transfer volumes. Be sure to stabilize the pH in the case of this dye with a buffer.

References

1. Baum P, Fundel-Clemens K, Kreuz S, Kontermann RE, Weith A, Mennerich D, Rippmann JF (2009) Off-target analysis of control siRNA molecules reveals important differences in the cytokine profile and inflammation response of human fibroblasts. Oligonucleotides 20(1):17
2. Lin X, Ruan X, Anderson MG, Mcdowell JA, Kroeger PE, Fesik SW, Shen Y (2005) siRNA-mediated off-target gene silencing triggered by a 7 nt complementation. Nucleic Acids Res 33:4527–4535
3. Wang HW, Noland C, Siridechadilok B, Taylor DW, Ma E, Felderer K, Doudna JA, E. (2009) Nogales structural insights into RNA processing by the human RISC-loading complex Nat. Struct Mol Biol 16:1148–1153
4. Murie C, Barette C, Button J, LafanechÒre L, Nadon R (2015) Improving detection of rare biological events in high-throughput screens. J Biomol Screen 20(2):230–241
5. Persengiev SP, Zhu X, Green MR (2004) Non-specific, concentration-dependent stimulation and repression of mammalian gene expression by small interfering RNAs (siRNAs). RNA 10 (1):12–18
6. Hwanga H-W, Wentzelb EA, Mendell JT (2009) Cell–cell contact globally activates microRNA biogenesis. Proc Natl Acad Sci U S A 106 (17):7016–7021
7. König R et al (2007) A probability-based approach for the analysis of large-scale RNAi screens. Nat Method 4(10):847–849. doi: 10.1038/nmeth1089
8. Huang DW, Sherman BT, Lempicki RA (2009) Systematic and integrative analysis of large gene lists using DAVID bioinformatics resources. Nature 4(1):44–57. https://doi.org/10.1038/nprot.2008.211

Chapter 9

Reporter Gene Assays Using Viral Functional Genomics Libraries

Genevieve Welch, Robert Damoiseaux, and Loren Miraglia

Abstract

While transfectable libraries are the workhorse for many screening cores, there is one obvious area where these reagents are not useful-hard to transfect cell lines and primary cells. One solution to this problem is the use of virus to introduce genomic reagents. This strategy is more commonplace now than ever before with libraries covering cDNAs, shDNAs, miRNAs, and guide RNAs readily available. Maintenance and use of these libraries are more challenging than the transient transfection approach due to the viral production step, and the infrastructure necessary to generate them. The following pages will delve into the details for working with arrayed well formats for both lentiviral and retroviral libraries.

Key words Functional genomics, Lentiviral, Retroviral, Gain of function, Loss of function, Library management, cDNA, shRNA, miRNA, CRISPR

1 Introduction

The first published large (>5000 genes) arrayed well lentiviral shDNA screen appeared in 2004 [1]. This was the beginning of a wave [2, 3] which continues today where retroviral and lentiviral libraries can be utilized for almost any cell line or cell type that is actively dividing (retrovirus—[4, 5]) or those that are both dividing and nondividing (lentivirus). A general overview of the life cycle of the production and usage of a (lenti)viral library is given in Fig. 1. The advantages go beyond the ability to transduce cells, which in itself is significant. When compared to transient transfection of cDNAs, siRNAs, and miRNAs, the viral introduction has the added benefit of sustained delivery after integration into the host cell genome. This allows for selection of cells that have been transduced by either virus containing fluorescent markers or antibiotic genes to yield a pure population of cells, all expressing the reagent of choice. Furthermore where transient transfection of cDNA or miRNA results in cells peaking in activity of the delivered reagent around 48 h and siRNA at 72–96 h, a virally transduced library will

Robert Damoiseaux and Samuel Hasson (eds.), *Reporter Gene Assays: Methods and Protocols*, Methods in Molecular Biology, vol. 1755, https://doi.org/10.1007/978-1-4939-7724-6_9,

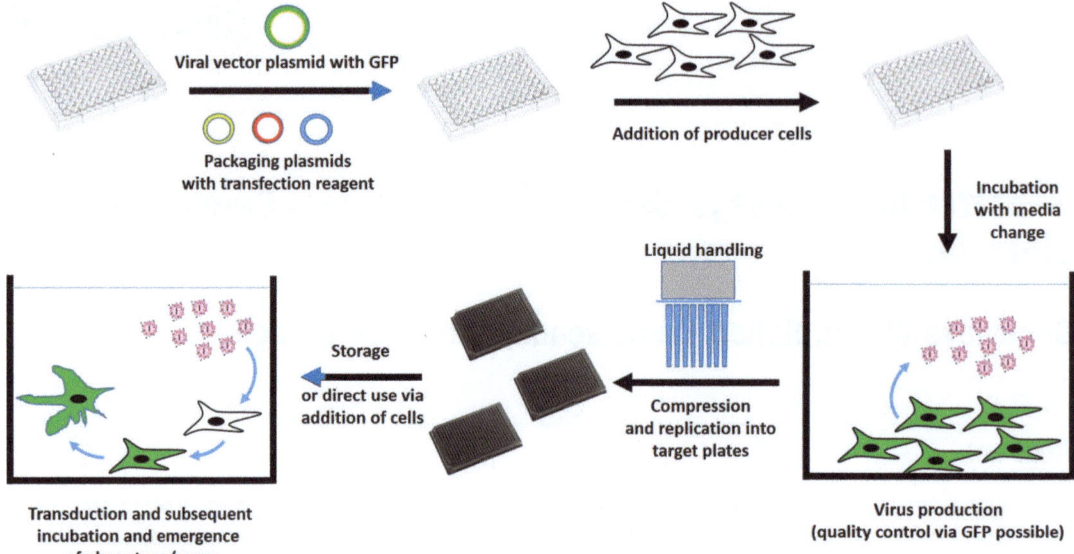

Fig. 1 Depiction of the viral production and transduction workflow. Blue arrowheads denote the possibility of breaks without degradation of performance and black arrowheads denote time-sensitive steps that need to be carried out immediately. The strict adherence to these breakpoints is critical for the proper functioning of the screening campaign. Note that there are only two breakpoints in a workflow with significant complexity and as such, the project has a premium on automation with proper scheduling capabilities

yield cells which can be utilized over much longer periods of time (potentially weeks). This type of approach is especially suitable for situations where the phenotype is developed over time since dilution of the delivered viral reagent does not occur as it does in transiently transfected cells.

While the advantages are many for using lentiviral and retroviral libraries, only one of these types-lentivirus-can be found from commercial vendors in an arrayed well format. This likely stems from the facts that retroviral cDNA vectors are more challenging to process from DNA through viral production, can also be prone to instability in the parental vector and the cost for cloning ORFs into an arrayed well format is considerable. However, the retroviral approach does have the distinct advantage of the ability to package larger inserts, which makes it preferable for cDNA overexpression. It is worth revisiting the other area of difference previously mention, the ability to transduce nondividing cells.

While these two types of viral delivery systems are similar with lentiviruses representing a subtype of retroviruses, only one lentivirus can integrate consistently into the genome of nondividing cells. This difference should be considered immediately upon project initiation as it will influence the cell type or line used for the assays. For example, both lentivirus and retrovirus can easily infect HEK293T cells, a SV40 transformed embryonic kidney line. However, when one works with a nondividing cell type such as adipocytes or myocytes, gammaretroviruses such as the commonly used murine leukemia virus (MuLV) are unable to produce infection due to the requirement for nuclear envelope breakdown for integration [6, 7]. This observation will be a decision point for identifying the desired approach, lentivirus or retrovirus, would be suitable for delivery based on the cells or cell types of the underlying biology. However, given that that the prevalence of lentiviral libraries far exceeds that of retroviral libraries, the majority of the focus for this section is on the handling, production, and use of lentiviral libraries with a small portion on retrovirus.

Finally, safety should always be a consideration when working with lentivirus or retrovirus. Current viral systems go far in separating the necessary components into multiple plasmids to significantly reduce the risk of attaining replication competency. For lentivirus, there are three plasmid and four plasmid production systems available. For retrovirus, most protocols utilize two or three plasmid systems where the viral envelope is on a separate plasmid from the cDNA. However, an important consideration with the introduction of a cDNA is that oncogenic transformation could occur. This same consideration can exist for shDNA inserts in that there is always the possibility that a gene knockout could result in an oncogenic transformation. For these reasons, it is highly recommended to utilize BSL 2+ practices and to confer with local safety regulations along with studying NIH biosafety protocols and information from the Centers for Disease Control before initiating any work.

2 Materials and Methods

2.1 General Considerations for Viral Libraries

Selecting a lentiviral shDNA library. When considering the appropriate lentiviral library for screening there are several important questions to ask including a marker of choice, the depth and coverage of the library, the availability of automation to process the library and cost. Each of these questions can steer the researcher toward one product or another. Carefully reviewing the requirements of both the researcher/screening group and the available libraries will ensure that there is a match. In the following pages, each question is reviewed in depth to provide sufficient information for the researcher.

Marker of choice. This first consideration will impact the type of protocols created for screening. The two most common markers found on commercial libraries are Green Fluorescent Protein (GFP) and Puromycin (Puro). Each marker has its advantages and disadvantages. GFP works well in FACS based and high content imaging screens. Selection can take place without antibiotic treatment and in an arrayed format, transduced cells can both be compared to untransduced in the same well and separated from them to determine if a phenotype can be associated to the gene knockout. However, in the case that there is a majority of untransduced cells, the confidence level drops. With the use of puromycin, cells which are not transduced simply die in the presence of the drug. The result is a pure population of cells with little or no background. In an arrayed well format, one must also consider that the readout can be affected by the number of cells remaining in each well. The disadvantage of this approach is that some cell lines cannot be treated with puromycin or are impacted to the extent that the biology is affected.

Depth and coverage. These attributes are similar if not identical to the siRNA screens. One must consider the gene families to be included in the library and the number of constructs per gene present in the library. Optimally this number will be 4–6 shDNA constructs per gene. Libraries larger than this size will add considerable cost to virus generation.

Availability of automation and cost. These two parameters are far and away the most critical to address when choosing a library. For an arrayed well lentiviral screen, one would need the following instrumentation: deep well block shaker, automated 96-well DNA purification robot and automation for normalization, liquid handler for DNA spotting, and a plate washer/dispenser for transient transfection/viral production. The same liquid handler for DNA spotting can then be used for transfer of the arrayed well virus to a 384-well destination plate containing the target cells. In the absence of this instrumentation, one can consider a pooled library. The cost of generating a pooled library is significantly less than an arrayed well library. This cost savings is due to the protocols employed to generate the content for each type of library. An arrayed well library requires cloning of each individual sequence from annealed primers into a host vector while sequences are synthesized on a chip and cloned en masse into a host vector for a pooled library. Both approaches then transform bacteria to amplify, but again, the arrayed well approach requires a well by well growth and extraction protocol versus a bulk transformation and extraction for the pool. These differences are extended into the production of viral particles where the cost of the well by well protocol is in the vicinity of $1.00/well while the DNA can be bulk transfected into host cells along with packaging plasmids to produce virus representing the complete library with a greater than 10-fold reduction in cost over the arrayed format. With the advent of NexGen sequencing, representation of the individual clones can be determined easily from pooled DNA [8]. In addition to these advantages, pooled screens are also preferable when there is a limited amount of targeted cells available for transduction, as would be the case when primary cells are harvested from mice. However, most screens utilizing functional genomics draw their strength from the ability to interface with high content screening, i.e., automated microscopy coupled with image recognition software. Prime examples are disease phenotypes there the mode of action of a given phenotype is not clearly understood and functional genomics is a means by which to decipher potential points of intervention and isolate disease modifying targets. Such approaches are only viable in an arrayed format and as such necessitate the use of a strong support infrastructure as discussed here.

2.2 Viral Transduction Optimization

Prior to screen initiation, thought must be put into the cell line of choice, and what effect transduction has on the biology of interest. The first step in this process is to identify the cell line, which will yield the desired phenotype. Characterization of the phenotype is key before examining transducibility. An example of this would be staining for a particular protein in a black clear bottom plate (imaging endpoint) or, as another example, the expression of a particular receptor, as characterized by FACS. In either case, once the expected readout is confirmed, the cells must be assessed for the ability to be infected by lentivirus. Pretitered virus prepared from the parental transfer vector is essential for this exercise, taking advantage of either a fluorescent marker (GFP, RFP) or selectable marker such as puromycin. For selectable markers, a kill curve is necessary to predetermine the appropriate amount of drug necessary to cause complete cell death when using a new cell type/line. Below is a minimal protocol for generating a kill curve with puromycin (*see* **Note 1**).

2.2.1 Identifying the Appropriate Concentration of Puromycin for a New Cell Line

1. Plate cells in a range of 0.1–1.0 $\times$ 10^4 cells per well into a 384-well plate (*see* **Note 2**).
2. Incubate overnight.
3. The following day add a titration of 500–10,000 ng/mL puromycin to selected wells leaving control wells untreated (*see* **Note 3**).
4. Replace media containing puromycin every 3 days over 6–10 days.
5. Remove media, wash with PBS and trypsinize cells.
6. Resupsend in FACS media.
7. Run FACS in order to evaluate the number of live cells to determine minimum concentration which causes complete cell death. Compare all treated samples to untreated control (*see* **Note 4**).

For transductions using lentivirus enhancers are commonly utilized. These reagents range from polybrene, protamine sulfate, retronectin, to DEAE Dextran. Polybrene, or hexadimethrine bromide, is a cationic polymer that improves transduction efficiencies and is found in many transduction protocols. It can be toxic though, and it is recommended that the reagent is tested for both the ability to improve transduction and toxicity so the appropriate dose can be identified. After incubation with any of these reagents, it is recommended that the media is replaced 10–24 h post delivery of the virus. This ensures integration of the virus and optimal growth conditions for the cells while virus is being produced. Longer incubation periods without a media exchange can result in significant toxicity to the producer cells. Below is a sample protocol using puromycin and limited dilution to determine transduction efficiency.

Reagent List

Polybrene (Hexadimethrine bromide) 2 mg/mL stock (Sigma # H9268).

Puromycin 10 mg/mL stock (Sigma #P9620).

384-well black clear bottom plate (Greiner #781986).

Target cells with appropriate growth media.

Premade positive control lentiviral particle (Sigma #SHC003).

1.7 mL Eppendorf tubes.

1. Plate cells in a 384-well black clear bottom plate at 50 μL at such a concentration that the confluency will be 30–50% the following day. Incubate 18–20 h at 37 °C with 5–7% CO_2.
2. Thaw out tube of unconcentrated lentiviral SHC003 stock.
3. Prepare tenfold serial dilutions of virus in Eppendorf tubes from 1×10^{-2} to 1×10^{-6}.
4. Prepare 1 mL of media with Polybrene at a final concentration of 8 μg/mL.
5. Remove 25 μL media from each well and add 12.5 μL of virus and 12.5 μL of the Polybrene mix.
6. Incubate for 24 h.
7. Remove 45 μL of media and replace with 45 μL of fresh media. Wells can be visualized via high content imaging 48–72 hours post transduction for % transduction (*see* **Note 5**).

2.3 Arrayed Well Lentiviral Library Preparation

The production of arrayed libraries is split into several segments. In the first segment, the DNA containing the shRNA vectors is spotted out together with their respective controls. The controls can also be added later with the media using devices that allow for direct access to individual wells such as noncontact liquid dispensers.

2.3.1 Spotting Protocol for Transducable Libraries (Transfer Vectors Only)

1. Select the plate types and number of copies to be stamped.
2. Barcode the 96-well plates with barcodes containing date, ID, and a reference to the plate map in the inventory.
3. Validate the precision and repeatability of the spotting.
4. Thaw out only the number of plates and reagents that can be processed in a single run.
5. Rearray wells to a working copy if possible leaving control wells empty. At this point the spotting process could be interrupted if the library plates are frozen down.
6. Spot the copies at a concentration of 100 ng/well. With plates normalized at 40 ng/μL, the spotting volume is 2.5 μL/well. This should be performed using an automated liquid handling system that scans barcodes and generates a trace for each pipetting transfer. This will aid with the recovery in case any irregularities are ever found.
7. Seal the copies with a plate seal that has been validated.
8. If production is expected within the next 24 h, plates can be placed at 4 °C. Plates must be spun the following day before initiating the transfection protocol to ensure that the total volume is available for mixing with packing plasmid transfection mix.

2.4 Lentiviral Production Protocol

In the next step of the production, the packaging plasmids are added together with the transfection reagent. The DNA–liposome complex is then allowed to form. This is concluded with the addition of a producer cell suspension. The next day, the media is removed and on the following day the virus can be collected. Care is needed in the selection of the producer cells and thorough optimization of the transfection process. Below is a standard viral production protocol (*see* **Note 6**).

Reagent List

T175 flasks.

Hek 293T cells.

Source DNA plates normalized to 40 ng/μL lentiviral shDNA per well in 96-well plate.

FuGENE® 6 Transfection Reagent (Promega).

Plasmid expressing Gag and Pol such as pMDLg/pRRE.

Plasmid expressing Rev such as pRSC-Rev.

Plasmid expressing envelope protein such as pVSV-G.

96-well viral producer plate (*see* **Note 7**).

HEK293T growth media (DMEM, high glucose, 10% FBS, 1× Pen/Strep/Glutamine, 1× nonessential amino acids, 1× HEPES, 1× sodium pyruvate).

HEK293T 2× growth media.

96-well V-bottom plate (Greiner #651261).

2.4.1 Transient Reverse Transfection of 293T Cells with Lentiviral Packaging and Transfer Vectors

1. Seed low passage number Hek293T cells (~18 million) in 10% FBS DMEM to a T175 flask so that the cell density will be approximately 80% confluent the next day.
2. Incubate at 37 °C, 5% CO_2, 95% humidity for 24 h.
3. Prepare transfection cocktail according to the following ratios (per well) in sterile glassware:
   (a) 50 μL serum free media (DMEM).
   (b) 0.87 μL FuGENE® 6.
   (c) 83 ng plasmid containing gag/pol.
   (d) 32 ng plasmid containing Rev.
   (e) 45 ng envelope plasmid.
4. Generate master mix by first adding DNAs to serum free media and then the appropriate amount of FuGENE® 6.
5. Mix by shaking vessel thoroughly.
6. Dispense 50 μL of mixture into each well of 96-well producer plates.
7. Incubate for 30 min.
8. While transfection is incubating, prepare HEK293T producer cells.
9. Harvest HEK 293T cells from flasks and resuspend at 8 × 10e5 cells/mL in 2× HEK293T growth media.
10. Dispense cell suspension at 50 μL/well for a final of 40,000 cells/well in 1× growth media.
11. Return plates to incubator at 37 °C, 5% CO_2, 95% humidity for 24 h.
12. Exchange media replacing with 70 μL of DMEM 10% FBS, leaving 30 μL behind in order to not disturb the cells from the bottom of the plates.
13. Return to incubator for an additional 24 h.
14. Transfer 90 μL from each well of the producer plates to the corresponding 96-well V-bottom plate for freezing and storage prior to transduction. In the cases where more than 90 μL are needed for the assay, replicate producer plates can be pooled into a single V-bottom plate.

Additional protocols for arrayed well lentiviral production can also be found in the literature [9–13].

2.5 Viral Transduction

The researcher has a choice of a reverse or forward transduction for lentiviral assays. The decision to choose one over the other is a function of multiple parameters. These parameters include—but are not limited to—the ability to transduce the target cell line in either forward or reverse format, the requirement that the target cell line be grown in well plates beforehand, and the automation processes involved in viral production. We have found instances where the target cell line is preferentially transduced in the forward format and have performed the screens in that format (unpublished data). However, the majority of the viral screens in our screening core utilize a reverse transduction protocol.

It almost goes without mentioning that target cells/cell lines should be well characterized before initiation of the screen. This includes cell line authentication as well as kill curves, cell density optimization after infection in addition to the standard assay validation that is necessary. The virus containing V-bottom plates can be thawed and spun before the proper amounts of viral supernatant are transferred using a liquid handler into destination 384-well plates. Transduction volumes usually range between 5 and 40 μL per well. Limitations on transduction volume are defined by both the transducibility of the target cell line and the acceptable volume in the 384-well plate. Transductions should be performed in replicate in order to perform advanced datamining (*see* **Note 8**).

2.6 Retroviral cDNA libraries

As mentioned in the introduction, arrayed well retroviral cDNA libraries are not available from vendors. However, one can purchase retroviral cDNA from multiple sources such as Clontech and Agilent Technologies/Stratagene. The main consideration for packaging is tropism. There are four different envelopes readily available, Eco, Ampho, 10A1, and VSVg. The Eco envelope is used for mouse and rat cells but not human. The Ampho envelope protein transduces most mammalian cells but not hamster. The 10A1 envelope is used to transduce most mammalian cells including human. Finally, the VSVg envelope protein generally has broad tropism. In some cases, the envelope is provided by a packaging cell line or the envelope can be provided through the packaging vector itself. Most of the basic retroviral vectors utilize wild-type LTRs from Moloney murine leukemia virus (MMLV) or from Murine Stem Cell Virus (MSCV) to ensure proper packaging and viral particle production. In the example below, the target cells are mouse in origin, and hence the use of the pCL-Eco envelope construct.

2.6.1 Viral Transduction Optimization

Please see section 2.2 for details. In general, a stock of titered virus, preferably generated from a parental transfer plasmid with a marker, should be available to assess transduction efficiencies of target cells.

2.6.2 Retroviral Library Production

Reagent List

T175 flasks.

Hek 293T cells.

Source DNA plates normalized to 20 ng/μL retroviral cDNA per well in 96-well plate.

FuGENE® 6 Transfection Reagent (Promega).

Packaging plasmid pCL-Eco (for mouse cell line).

96-well viral producer plate (Greiner #655980).

384-well black clear bottom plate for imaging (Greiner # 781986).

DMEM, high glucose.

DMEM 20% FBS.

For 12 × 96-well producer plates

1. Seed low passage number Hek293T cells (~18 million) in 10% FBS DMEM to a T175 flask so that the cell density will be approximately 80% confluent the next day. Prepare 1 flask for every 4 96-well producer plates.
2. Incubate at 37 °C, 5%CO_2, 95% humidity for 24 h.
3. Spot 1 μL of arrayed retroviral cDNA per well into 12 × 96-well producer plates This can also be done the day before and plates can be stored at 4 °C.
4. Harvest HEK293T cells and resuspend at 8 × 10e5 cells/mL in HEK293T 2× growth media.
5. Make transfection mix with 50 μL serum free media and 0.26 μL of FuGENE® 6 per well.
6. Add packaging plasmid pCL-Eco at 60 ng/well to the master transfection mix.
7. Dispense 50 μL/well of the master mix using dispenser.
8. Incubate for 30–60 min at room termperature.
9. Dispense 50 μL/well HEK293T cell suspension and return plates to plate incubator at 37 °C, 5%CO_2, 95% humidity for 24 h.
10. Exchange media replacing with 70 μL of DMEM 10% FBS, leaving 30 μL behind in order to not disturb the cells from the bottom of the plates.
11. Return to incubator for an additional 24 h.
12. Transfer 90 μL from each well of the producer plates to the corresponding 96-well V-bottom plate for freezing and storage prior to transduction. In the cases where more than 90 μL are needed for the assay, replicate producer plates can be pooled into a single V-bottom plate.

3 Notes

1. All initial experiments such as virus titrations and kill curves with antibiotics should be performed as 1:1 dilutions in triplicate.
2. A cell density titration is standard procedure before any assay is established. This is to include titrations of any additional variables such as antibiotics/transduction aids like polybrene. High content screening modalities are useful for making this work intensive part of the assay optimization easier.
3. By using a control well with no antibiotic along with a wide range, the researcher can identify ineffective or low doses (minimal toxicity), an optimal dose which is the lowest concentration where all the cells are dead and high doses where cells are dead within the 2–3 day range.
4. Flow cytometry is a very useful modality for smaller scale experiments such as this but can be limiting in throughput. High content screening is a powerful alternative where Hoechst 33342 with Propidium iodide at standard concentration can be employed. However, some cells like stem cells—especially hematopoetic stem cells—can exclude Hoechst.
5. Some cell types may also show other effects of viral infection such as senescence. It is recommended that the researcher be on the lookout for changed or unusual cell morphology.
6. The plasmids pMDLg/pRRE and pRSC-Rev are part of the third generation lentiviral packaging system generated in the Didier Trono lab [12] while the plasmid pCMV-VSV-G was created in the Bob Weinberg lab [13]. The third generation system is the standard for safety in that all unnecessary elements have been removed from the plasmids and the separation of the constituent parts to four plasmids (Gag/Pol, RRE, VSV-G, transfer plasmid) minimized the possibility to generate replication competent virus. Specifically, the viral protein Tat has been removed and replaced with a hybrid LTR/promoter, making it Tat independent. It is strongly recommended to review the safety precautions of lentiviral production before commencing experiments.
7. The choice of plate type for viral production is also a critical parameter worth discussing in detail for several reasons. First, for all our viral production, we use HEK293T cells. These cells are the workhorse for many labs in regard to viral production. One characteristic that they have is the propensity to come off the plate surface, especially when challenged with the task of viral production. For this reason, we use the Advanced TC surface from Greiner [14]. There are products from other

vendors that offer a similar advantage, i.e., a modified surface which increases binding of the cells to the bottom of the well. The ability to remain attached improves all facets of the production cycle from cell viability to gene expression and ultimately higher viral titer.

8. It is imperative that large production runs undergo rigorous quality control before screening transductions are initiated. There are multiple ways in which this can be achieved, but in almost all cases either an antibiotic or fluorescent marker is utilized. In order to perform quality control, one must freeze down the production plates while the process takes place. It is recommended that for every ten viral production plates produced, 15 μL of a single row (for example H01–H12) is removed. This virus can be frozen overnight at −80 °C (to mimic the screening transduction conditions) and then used for quality control by thawing and then transducing target cells in 384-well format, with one set of wells transduced with 10 μL, and a second set with 1 μL. For viruses which contain an antibiotic, the conditions determined from the optimization process are used (identified dose of antibiotic, identified treatment regimen). Transduction efficiencies can be determined by FACS by comparing antibiotic treated cells with untreated cells and compared to results identified from the optimization process (expected percentage of live cells after antibiotic treatment). For fluorescent markers, the process takes only 2 days, using the same approach, by either FACS or high content imaging to determine the percentage of cells successfully transduced. This again can be compared to the results derived from the optimization process.

References

1. Berns K, Hijmans EM, Mullenders J, Brummelkamp TR, Velds A, Heimerikx M, Kerkhoven RM, Madiredjo M, Nijkamp W, Weigelt B, Agami R, Ge W, Cavet G, Linsley PS, Beijersbergen RL, Bernards R (2004) A large-scale RNAi screen in human cells identifies new components of the p53 pathway. Nature 428 (6981):431–437
2. Moffat J, Grueneberg DA, Yang X, Kim SY, Kloepfer AM, Hinkle G, Piqani B, Eisenhaure TM, Luo B, Grenier JK, Carpenter AE, Foo SY, Stewart SA, Stockwell BR, Hacohen N, Hahn WC, Lander ES, Sabatini DM, Root DE (2006) A lentiviral RNAi library for human and mouse genes applied to an arrayed viral high-content screen. Cell 124 (6):1283–1298
3. Klinghoffer RA, Roberts B, Annis J, Frazier J, Lewis P, Linsley PS, Cleary MA (2008) An optimized lentivirus-mediated RNAi screen reveals kinase modulators of kinesin-5 inhibitor sensitivity. Assay Drug Dev Technol 6 (1):105–119
4. Springett GM, Moen RC, Anderson S, Blaese RM, Anderson WF (1989) Infection efficiency of T lymphocytes with amphotropic retroviral vectors is cell cycle dependent. J Virol 63 (9):3865–3869
5. Miller DG, Adam MA, Miller AD (1990) Gene transfer by retrovirus vectors occurs only in cells that are actively replicating at the time of infection. Mol Cell Biol 10(8):4239–4242
6. Lewis PF, Emerman M, Virol J (1994) Passage through mitosis is required for

oncoretroviruses but not for the human immunodeficiency virus. J Virol 68(1):510–516

7. Yamashita M, Emerman M (2006) Retroviral infection of non-dividing cells: old and new perspectives. Virology 344(1):88–93
8. Sims D, Mendes-Pereira AM, Frankum J, Burgess D, Cerone MA, Lombardelli C, Mitsopoulos C, Hakas J, Murugaesu N, Isacke CM, Fenwick K, Assiotis I, Kozarewa I, Zvelebil M, Ashworth A, Lord CJ (2011) High-throughput RNA interference screening using pooled shRNA libraries and next generation sequencing. Genome Biol 12(10):R104
9. Shum D, Djaballah H (2014) Plasmid-based shRNA lentiviral particle production for RNAi applications. J Biomol Screen 19 (9):1309–1313
10. Rines DR, Tu B, Miraglia L, Welch GL, Zhang J, Hull MV, Orth AP, Chanda SK (2006) High-content screening of functional genomic libraries. Methods Enzymol 414:530–565
11. Genetic Pertubation Platform (2017) Broad Institute. http://portals.broadinstitute.org/gpp/public/resources/protocols. Accessed 26 May 2017
12. Dull T, Zufferey R, Kelly M, Mandel RJ, Nguyen M, Trono D, Naldini L (1998) A third-generation lentivirus vector with a conditional packaging system. J Virol 72 (11):8463–8471
13. Stewart SA, Dykxhoorn DM, Palliser D, Mizuno H, EY Y, An DS, Sabatini DM, Chen IS, Hahn WC, Sharp PA, Weinberg RA, Novina CD (2003) Lentivirus-delivered stable gene silencing by RNAi in primary cells. RNA 9 (4):493–501
14. (2011) Advanced TC^{TM}: An innovative surface improving cellular assays. In: Forum-Technical notes and application for laboratory work. Greiner Bio-One International. https://shop.gbo.com/en/row/articles/catalogue/articles/0110_0110_0060/. Accessed 26 May 2017

Chapter 10

Using YFP as a Reporter of Gene Expression in the Green Alga *Chlamydomonas reinhardtii*

Crysten E. Blaby-Haas, M. Dudley Page, and Sabeeha S. Merchant

Abstract

The unicellular green alga *Chlamydomonas reinhardtii* is a valuable experimental system in plant biology for studying metal homeostasis. Analyzing transcriptional regulation with promoter-fusion constructs in *C. reinhardtii* is a powerful method for connecting metal-responsive regulation with *cis*-regulatory elements, but overcoming expression-level variability between transformants and optimizing experimental conditions can be laborious. Here, we provide detailed protocols for the high-throughput cultivation of *C. reinhardtii* and assaying Venus fluorescence as a reporter for promoter activity. We also describe procedural considerations for relating metal supply to transcriptional activity.

Key words Algae, Iron, Zinc, Copper, Yellow fluorescent protein

1 Introduction

Originally isolated from the jellyfish *Aequorea victoria* [1, 2], green fluorescent protein (GFP) was quickly adopted as a reporter in multiple organisms including bacteria [3, 4], plants [5, 6], and animals [3, 7]. The widespread use of GFP and its variants (fluorescent proteins (FPs) engineered to have different physical and spectral properties [8]) is due in large part to their ease of use. FPs derived from GFP form their chromophore autocatalytically [3]; detection of fluorescence in vivo does not require the addition of a substrate and does not necessarily require sample processing. The most common applications of FPs include localization of FP-protein fusions in vivo and gene expression studies of promoter-*FP* fusions. In the first case, the FP serves as a fluorescent label for proteins of interest; in the second case, the magnitude of fluorescence is a proxy for promoter activity (Fig. 1).

Here, we describe the use of promoter-*YFP* fusions to profile promoter activity in response to metal availability in the green alga *Chlamydomonas reinhardtii*. YFP (Yellow Fluorescent Protein) is a general term used to refer to variants of GFP that carry amino acid

Robert Damoiseaux and Samuel Hasson (eds.), *Reporter Gene Assays: Methods and Protocols*, Methods in Molecular Biology, vol. 1755, https://doi.org/10.1007/978-1-4939-7724-6_10,

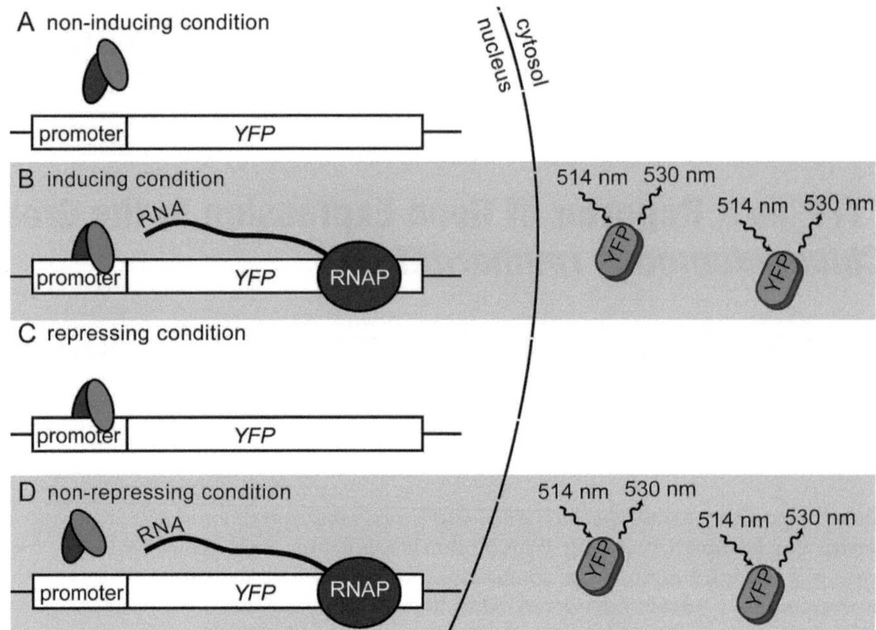

Fig. 1 Diagram of promoter-*YFP* fusions. In panels A and B, expression of YFP is dependent on a promoter fragment that contains a binding site for a transcriptional inducer. When transferred to inducing medium the transcription factor binds to the promoter fragment, transcription occurs, and YFP accumulates in the cytosol. If the binding site for an inducer is deleted, YFP will no longer be expressed in the inducing medium. In panels C and D, expression of YFP is dependent on a promoter fragment that contains a binding site for a transcriptional repressor. In non-repressing medium, the transcription factor no longer binds to the promoter fragment, transcription is allowed to occur, and YFP accumulates in the cytosol. If the binding site for a repressor is deleted, YFP will be expressed in both media. In the protocol, the medium that leads to YFP expression is referred to as the "inducing medium" without an assumption of the regulatory mechanism. Abbreviations: RNAP, RNA polymerase; YFP, yellow fluorescent protein

substitutions that shift the protein to a yellowish emission (emission peak at 530 nm) [8, 9]. This protocol specifically employs Venus, an engineered YFP resulting from several amino acid substitutions that provide increased fluorescence intensity, faster maturation, and relative tolerance to pH and chloride ions as compared to its predecessor (EYFP, enhanced yellow fluorescent protein) [10]. The exploitation of FPs, specifically YFP, in *C. reinhardtii* has been facilitated in recent years by the availability of constructs and strains developed to overcome inherently poor protein expression associated with non-native genes expressed from the nuclear genome [11]. The robustness of FP-based systems in *C. reinhardtii* (and thus the ability to use these reporter systems with promoters of lowly expressed genes) continues to improve [12–14].

The YFP reporter protocol described here has several attractive qualities. Individual strains are arrayed in 96-well format, and fluorescence is measured on a microplate reader (such as a Tecan Infinite® M1000 or Molecular Devices SpectraMax®), enabling multiple independent transformants to be assayed simultaneously (see additional information about replicates in Subheading 3.2). This is an important consideration for any reporter system in *C. reinhardtii*, as introduced DNA predominately recombines into the nuclear genome during nonhomologous end-joining, and the expression level of the promoter fusion varies between transformants in part because of the specific site of integration in each case [15–17] (*see* **Note 1** in section 4). Therefore, multiple independent transformants are essential for assessing the significance of variation in promoter fusion activity between conditions or between wild-type and mutated promoter regions. Samples are assayed for YFP activity without upstream sample processing. As a result, assay time equates to the speed of a plate reader, which is typically less than 5 min per plate. Because there is no need for sample processing, fluorescence of whole cultures grown in the presence and absence of stimuli can be measured over the course of the growth curve without removing cells. By using bottom-read fluorescence, which does not require removal of the plate lid, cultures are more easily maintained free of contamination. Because of these attributes, optimizing the growth conditions and optimal point in the growth curve to measure YFP fluorescence for a subsequent mutational analysis (i.e., largest difference between WT promoter-*YFP* fluorescence from test and control conditions, *see* Fig. 3c for an example) is straightforward and relatively simple.

Since *C. reinhardtii* is a valuable single-celled reference organism for understanding metal homeostasis in the plant lineage [18–20], we also provide details for growing cultures with controlled metal nutrition. Because many metal ions are nutrients and potential toxins, plants must fine-tune morphological, physiological, and molecular responses to meet the catalytic demand for these elements while avoiding toxicity. How the cell achieves this balance through gene regulation is an active field of investigation. For studies of metal-responsive gene regulation, *C. reinhardtii* is an advantageous experimental system for several reasons [21, 22], including: (1) metal supply can be tightly controlled without the need to resort to using metal chelators, aiding reproducibility and avoiding misinterpretation of results due to the nonspecific nature of chelators, (2) the ability to distinguish between absence of a metal and presence of the metal but with reduced accessibility (a distinction that is missing in many papers in the literature), and (3) availability of transcriptomic datasets and well-characterized markers of iron, zinc, and copper deficiency [23–25]. To identify *cis*-acting regulatory sequences and the *trans*-acting metalloregulators in *C. reinhardtii*, we detail several considerations that should be made when correlating promoter activity with metal availability.

2 Materials, Equipment, and Experimental Considerations

1. *Chlamydomonas reinhardtii* str. UVM11; this strain has higher YFP expression from the nuclear genome than typical laboratory strains ([11], **strain UVM11 may be obtained from authors on agreement with a Material Transfer Agreement**).
2. DNA for transformation: (1) pJR39GW containing the promoter fusion construct to be analyzed (referred to in the protocol as the "test promoter"), (2) pJR39GW (or other appropriate negative control), and (3) pJR39GW containing a promoter fragment that can be used as a positive control, such as the *FOX1* promoter [26], *ZRT1* promoter [23], or *CYC6* promoter [27] as a positive control for iron-, zinc-, or copper-responsive *YFP* expression, respectively. pJR39GW is a Gateway-adapted version of pJR39 [11] and was constructed by replacing the *PsaD* promoter of pJR39 with a Gateway cassette. pJR39GW contains *aphVIII*, which encodes resistance to paromomycin for selection of *C. reinhardtii* transformants, *bla* for selection of *E. coli* transformants with ampicillin, and the Venus variant of YFP as the reporter protein. *See* (Chapter 11 Blaby and Blaby-Haas) in this series for details regarding the cloning of a promoter fragment into Gateway-adapted vectors.
3. Growth conditions for *C. reinhardtii*: Aseptic technique and clean, organized working practices should be applied at all stages. These practices are important to avoid contamination of cultures with both unwanted organisms and metals [28]. *C. reinhardtii* and sterile media should be handled in a laminar flow hood with presterilized culture plates, pipette tips, and pipettor (the barrel can be sprayed with 70% ethanol). Standard growth conditions are 23 °C, ~80 μmol m^{-2} s^{-1} with a ratio of 1 warm fluorescent light bulb (3000 K) to 2 cool fluorescent light bulbs (4100 K) and rotational shaking at 180 rpm, e.g., in Innova 44 incubators (Innova, New Brunswick Scientific, Edison, NJ), for liquid cultures, or stationary in growth chambers (Percival CU-36L6 or similar) for petri dishes and 96-well plates.
4. *C. reinhardtii* "metal-free" growth medium (as described in Quinn and Merchant [28]): Tris-acetate phosphate (TAP) medium [29] with high purity (>99.9% trace metals basis) chemicals (available from Sigma-Aldrich, ACROS organics or AlfaAesar; certificates of analysis are typically available before purchase of chemicals and can be used to estimate whether concentrations of contaminating metals in the prepared medium would be acceptable) and milliQ-H_2O: each liter of TAP contains 10 mL TAP salts solution (15 g NH_4Cl, 4 g $MgSO_4·7H_2O$, 2 g $CaCl_2·2H_2O$, final volume of 1 L milliQ-

grade H_2O; to avoid precipitation dissolve the $CaCl_2$ in 300 mL milliQ-grade H_2O and the NH_4Cl and $MgSO_4$ in 500 mL milliQ-grade H_2O; then mix the two solutions together and bring up to the final volume), 8.3 mL phosphate solution (18.5 g K_2HPO_4 and KOH to pH of 7.1 (about 28 mL of 20% KOH), final volume 1 L), 10 mL Tris-acetate solution (242 g Tris, 100 mL glacial acetic acid, final volume of 1 L milliQ-grade H_2O) and 1 mL each component of high-purity Kropat trace metal solutions [30]. Media stock solutions (TAP salts, phosphate and Tris-acetate solutions) and Kropat trace metal solutions are stored in metal-free plastic (to prevent metal contamination of solutions in the laboratory, storage containers for metal stock solutions should not be reused to make other solutions or washed with general-use laboratory glassware). Dedicated "metal-free" plastic graduated cylinders should be used. To make appropriate metal-deficient or -free medium, reduce or leave out, respectively, the necessary metal stock solution. When making TAP, use disposable plastic serological pipette tips and make media in freshly acid-washed glassware (rinsed three times with 6 N HCl followed by 7 rinses with milliQ-grade H_2O, the glassware should be used within 2 days or re-acid-washed). Media should be made no earlier than the day before (or morning of) 96-well plates are to be filled to prevent leaching of metals from the glass/plastic into the media.

5. Select agar (Thermo Fisher 30391-023), washed three times in milliQ-grade H_2O, is used at 1.5% w/v for solid growth medium.
6. Agar-solidified TAP in deep-dish petri dishes (Fisher Scientific FB0875711) containing 5 μg/mL paromomycin (Sigma P5057).
7. Light microscope (such as a ZEISS Primo Star; bright-field illumination is sufficient for counting *C. reinhardtii* cells).
8. Hemocytometer with cover slip (Sigma Z359629).
9. Low-adhesion microcentrifuge tubes, 2 mL (USA Scientific 1420-2600), filled to the 250–300 μL demarcation with glass-beads (particle size, 425–600 μm, acid-washed; Sigma G8772), autoclaved.
10. Sterile 50 mL and 15 mL conical polypropylene centrifuge tubes (Thermo Scientific 339652 and 339650, respectively).
11. Centrifuge and rotors appropriate for centrifuging 50 mL tubes (Thermo Scientific 339652) at 2000 × *g* and for 15 mL tubes (Thermo Scientific 339650) at 1000 × *g*.
12. Vortex mixer.

13. Sterile 96-well plates: for master plates can use clear plates with clear lids (such as the Nunc™ Edge 96-Well Plate, which has a moat for humidity control, Thermo Scientific 267313); for culture plates used for fluorescence measurements use black with clear, flat bottoms and clear lids (Thermo Scientific 165305).
14. Cell culture roller drum (Fisher Scientific 14-277-2) or tube rotator.
15. Multichannel pipette, 10–100 μL and 30–300 μL, and sterile pipette tips.
16. Presterilized multichannel basins (Fisher Scientific 13-681-500).
17. Plate reader capable of taking measurements of A_{750nm}, and bottom-read fluorescence with Emission (Em): 514 nm/Excitation (Ex): 530 nm and Em:440 nm/Ex:680 nm (such as a Tecan M1000 Pro).

3 Methods

3.1 Bombardment by Glass Beads—See Fig. 2 for a Workflow Schematic (Please Refer to Neupert et al. [31] for Additional Details Regarding Transformation and Kindle [32] for Original Protocol)

1. Using PsiI (or any restriction enzyme that cuts the plasmid outside of the promoter-*YFP* fusion and *aphVIII* region) linearize the negative control, positive control, and test promoter plasmids.
2. Grow a 300 mL liquid culture of UVM11 under standard growth conditions to a density of 1×10^6 cells mL^{-1}. Cell density is determined using a light microscope and a hemocytometer [33].
3. Transfer 50 mL of culture to sterile centrifuge tubes (e.g., 50 mL Falcon tubes).
4. Collect cells by centrifugation at $2000 \times g$ for 5 min, room temperature. Pour off the supernatant.
5. Gently resuspend the pellet in sterile TAP by slowly pipetting up and down with a 1 mL pipette tip to a final cell density of 4×10^8 cells mL^{-1}.
6. Transfer 200 μL aliquots of cells to a preautoclaved 2 mL low adhesion tube containing glass beads and DNA (500 ng linearized DNA in total volume of 5 μL).
7. Vortex at highest setting for 15 s.
8. The beads will settle quickly to the bottom of the tube. Gently collect the cells using a 100–1000 μL single-channel pipette and sterile pipette tip. Transfer cells to a 15 mL sterile centrifuge tube preloaded with 5 mL TAP.
9. Allow the cells to recover by incubating on a benchtop roller drum, $1 \times g$ (~50–60 rpm), for 8 h, in low light (<20 μmol

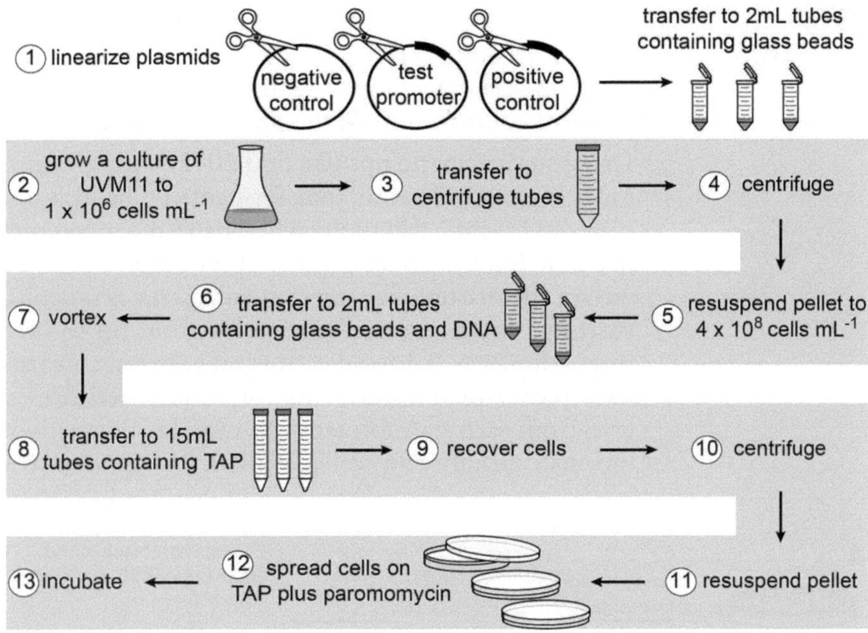

Fig. 2 An experimental workflow of protocol Subheading 3.1. Circled numbers correspond to steps in the protocol for UVM11 transformation by glass bead bombardment. Three plasmids—the first contains a promoter-less *YFP* construct (negative control), the second contains the promoter of interest upstream of *YFP* (test promoter) and the third contains a promoter-*YFP* fusion that is metal-responsive (positive control)—are linearized with a restriction enzyme. Prior to transformation in **step 6**, this DNA should be added to the preautoclaved glass beads. Since there are three *YFP* constructs, **steps 6–13** are carried out on three separate aliquots of cells

m^{-2} s^{-1}). The purpose of this step is to allow the cells to recover and express the resistance marker before selection, and it is important that the cells do not undergo a round of cell division to ensure that each colony is from an independent integrant.

10. Collect the cells by centrifugation at 1000 × *g* for 5 min, room temperature, and pour off the supernatant.
11. Gently resuspend the pellet by slowly pipetting up and down with a 1 mL pipette tip in the remaining volume of TAP left in the tube after decanting.
12. Gently spread cells with an ethanol-sterilized glass spreader or a presterilized plastic spreader onto agar-solidified TAP containing 5 μg/mL paromomycin.
13. Incubate in growth chamber under standard growth conditions until colonies are visible (~5–7 days).

3.2 YFP Assay—See Fig. 3a for a Workflow Schematic

1. Using a multichannel pipette with sterile tips, fill three 96-well plates with 200 μL TAP per well. These are the master plates for growing cells transformed with the (1) test promoter, (2) the negative control, and (3) the positive control.
2. Using sterile pipette tips (we find 20–200 μL size tips to be the most appropriate) array transformants by hand into a 96-well plate (use aseptic technique when handling, open only in laminar flow hood). To keep track of inoculation, leave tip in the well of the master plate until an entire row is inoculated. Then swirl tips to ensure transfer of cells from tip to medium and throw tips away. To avoid confusion between constructs and to have plenty of transformants for robust statistics (since YFP level from each transformant will vary due to positional effects) inoculate an entire 96-well plate with 96 individual colonies per

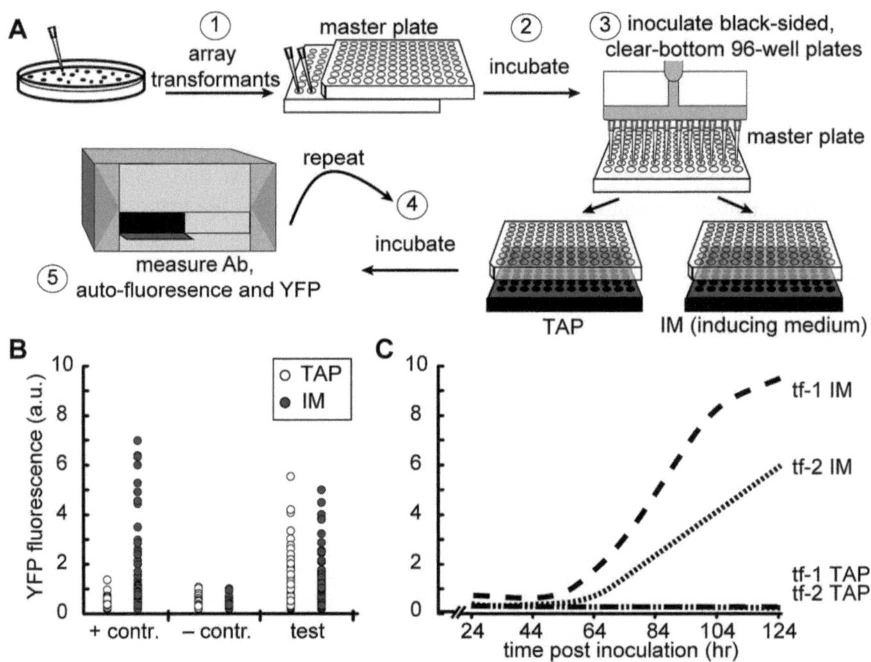

Fig. 3 An experimental workflow of protocol Subheading 3.2. (**a**) Circled numbers correspond to steps in protocol Subheading 3.2. Since there are three *YFP* constructs, **step 1** involves three petri dishes of transformants (corresponding to the negative control, test promoter and positive control) and three 96-well master plates filled with TAP. **Step 3** requires a total of 6 black-sided, clear-bottom plates; inducing medium is abbreviated as IM. (**b**) YFP fluorescence from 96 UVM11 cultures each transformed with a positive control (+ contr.; the native promoter for gene X), a negative control (– contr.; a construct where the necessary transcriptional activation sequences were removed), or a test promoter fusion construct (test; a construct with a deletion in a transcription factor binding site; the conclusion from this result is that the deletion of this site leads to loss of repression in TAP medium). The 192 transformants were grown in 96-well plates filled with TAP (TAP) and 96-well plates filled with inducing medium (IM) for 6.5 days. In this example, YFP fluorescence is normalized to chlorophyll fluorescence. (**c**) plot of YFP fluorescence normalized to chlorophyll fluorescence for two individual transformants (tf-1 and tf-2) containing a native gene promoter fragment fused to the YFP gene grown in TAP (TAP) and inducing medium (IM)

construct and label plate appropriately. Since variability due to the site of integration is substantially greater than technical variability (due to investigator handling, aspects of the protocol, or plate reader), it is advantageous to have independent transformants rather than technical replicates (repeated measurements of the same sample); for further discussion on experimental replicates, the reader is encouraged to refer to Blainey, Krzywinski, and Altman [34].

3. Place plates in growth chamber and incubate under standard growth conditions as described above for 5 days.
4. Using a multichannel pipette and sterile tips, fill three clear-bottom, black-sided 96-well plates with 200 μL TAP per well and three clear-bottom, black-sided 96-well plates with 200 μL TAP minus the metal of interest or TAP with a reduced concentration of the metal of interest per well. For the rest of the protocol, TAP with minus/reduced metal content will be referred to as "inducing medium."
5. Working with a single master plate at a time, start by resuspending the cells in row A by gently pipetting up and down with a multichannel pipette using sterile tips. Without removing tips from the multichannel pipette, transfer 10 μL from row A of the master plate to row A of a fresh 96-well plate (with black sides and clear bottoms to avoid fluorescent signal bleed-through between wells) containing TAP, then add 10 μL from row A of the master plate to row A of a second plate containing inducing medium. Change tips to prevent cross-contamination of wells and repeat with row B, then C and so on. Once finished inoculating from the first master plate, start with the second master plate, followed by the third master plate.
6. Incubate the 96-well plates in a growth chamber under standard growth conditions. Care should be taken to ensure that plates receive the same light intensity, thus ensuring similar growth rates between plates.
7. After 24 h measure the absorbance (750 nm), chlorophyll fluorescence (Ex: 440 nm, 9 nm bandwidth; Em: 680 nm, 20 nm bandwidth) and YFP fluorescence (Ex: 514 nm, 5 nm bandwidth; Em: 530 nm, 5 nm bandwidth) of each plate. Subsequent to reading, place the plates back in the growth chamber.
8. Repeat measurements every 24 h for 7 days. For fluorescence measurements, the gain should not be changed between measurements. We routinely use a gain of 50 for chlorophyll fluorescence and 150 for YFP (using a Tecan M1000 PRO; for other plate readers these parameters should be determined empirically). Two 96-well plates, one filled with sterile TAP

medium and one filled with sterile inducing medium should be used as blanks.

9. If the YFP fluorescence signal from the cells transformed with the test promoter in inducing medium is not significantly higher than from those same transformants grown in TAP medium after 4 days, perform a second round of metal depletion. The purpose of this step is to further deplete the cells of metal ions that had been transferred during inoculation from the master plate, as the carry over may be sufficient to suppress the activity of the metal-responsive promoter fragment (*see* Subheading 3.3, **step 3** for a note on positive controls). Using a multichannel pipette and presterilized filter tips, repeat **steps 4** and **5**, except that instead of inoculating fresh plates from the master plates, transfer ≤10 μL from the 96-well plate containing transformants grown in inducing medium into fresh, sterile inducing medium, and transfer the same volume from the 96-well plate containing transformants grown in TAP into fresh TAP.
10. Repeat **steps** 7 and **8**.

3.3 Data Analysis

1. *Normalization*: YFP fluorescence can be normalized to absorbance and/or chlorophyll fluorescence, which are used as proxies for cell density. Which normalization to use should be determined empirically by calculating the proportionality of absorbance and chlorophyll fluorescence as a function of cell number when cells are grown in TAP and inducing medium, i.e., make standard curves. If growth in inducing medium negatively impacts chlorophyll content (for example, cells grown with suboptimal iron nutrition are chlorotic compared to growth with sufficient iron nutrition [35]) then the amount of YFP per cell will be overestimated when normalized to chlorophyll fluorescence. Cell size can also change in response to metal nutrition (e.g., the average cell size is larger when *C. reinhardtii* is grown in the absence of zinc compared to the presence of zinc [23]), which will affect the absorbance measurement [36]. If both measurements are inappropriate for normalization between the two growth conditions, an option is to take a subset of transformants (such as the 3 transformants that appear to have the highest inducible YFP activity) and quantify the cell number using a hemocytometer and microscope. However, this measurement will reduce the volume of the corresponding wells by 20 μL and should only be done as an endpoint.
2. *Plotting data*: The normalized YFP fluorescence for each transformant from a single time point can be plotted for TAP and inducing medium as in Fig. 2b. Additionally, YFP fluorescence for each transformant can be plotted as a fold-change between

normalized YFP fluorescence measured from TAP and inducing medium. Time course data can also be plotted, but for ease of visualization (since 96 transformants in 2 growth conditions will generate 192 curves per construct) the normalized YFP fluorescence from 2 or 3 transformants grown in TAP and inducing medium can be plotted (Fig. 2c).

3. *Statistics, controls and troubleshooting*: Possible sources of false negative and false positive results include expression-level variability due to genome position effects of the YFP-construct, background fluorescence (detection of fluorescence with Ex: 514 nm/Em: 530 nm not due to YFP), and data normalization. Therefore, there are several ways to analyze the data to provide confident conclusions about the activity of the test promoter.
   (a) Is there a statistically significant difference between the normalized YFP fluorescence of test promoter transformants grown in TAP and inducing medium?
      - Calculate the probability associated with a Student's *t*-test. The convention in most biological research is to use a significance level of 0.05.

   (b) If the answer is *yes* to (a):
      - Is the average fold-change for the test promoter significantly different from the average fold-change for the negative control?
         - Calculate the fold-change in normalized YFP fluorescence between inducing medium and TAP for each transformant.
         - Calculate the probability that the average test-promoter fold-change is different from the average negative-control fold-change with a Student's *t*-test. The convention in most biological research is to use a significance level of 0.05.

   (c) If the answer is *no* to (a):
      - Is the average fold-change for the positive control significantly different from the average fold-change for the negative control?
         - Calculate the fold-change in normalized YFP fluorescence between inducing medium and TAP for each transformant.
         - Calculate the probability that the average positive-control fold-change is different from the average negative-control fold-change with a Student's *t*-test. The convention in most biological research is to use a significance level of 0.05.

– The normalized YFP fluorescence from the positive control transformants should be significantly different from the negative control. If not, then there was either contaminating metal in the inducing medium leading to repression of the promoter construct or absence of induction, or technical issues with the plate reader. Differentiating between technical issues due to the growth conditions or plate reader can be tested for by repeating this protocol with a constitutively expressed YFP construct (such as pJR39 [11]). The resulting transformants should have statistically significantly more YFP fluorescence when grown in TAP as compared to the negative control.

4 Notes

1. Recently, the use of CRISPR/Cas9 to engineer the *C. reinhardtii* nuclear genome has been met with some success [41, 42]. However, this technique is still in its infancy and has not been applied to the creation of a genomic "landing site" for specific integration of the promoter fusion construct, which may aid in reducing the observed transformant-to-transformant expression variability.
2. YFP has a relatively long half-life [37]. Therefore, if time-resolved down-regulation of promoter activity is to be studied, the reader is pointed to the luciferase reporter gene system [38–40]. Luciferase has a relatively short half-life (*Renilla reniformis* luciferase expressed in *C. reinhardtii* and targeted to the cytosol has a half-life estimated to be 2 h [38]). In theory, if promoter activity of the reporter fusion is repressed, less time is required to observe an equivalent repression of luciferase activity compared to YFP activity, however, these two reporter systems have yet to be examined side-by-side in *C. reinhardtii.*
3. Unfortunately, not all promoters have strong enough expression levels for detectable YFP fluorescence with this system. Recently, several codon-optimized YFP constructs have been published that may aid in studying these types of promoters [13, 14].

Acknowledgments

This work was supported by the Division of Chemical Sciences, Geosciences, and Biosciences, Office of Basic Energy Sciences of the US Department of Energy (DE-FD02-04ER15529) and the National Institutes of Health (NIH) R24 GM42143 to SM, and the Office of Biological and Environmental Research of the United States Department of Energy (CEB-H). We are grateful to Prof. Ralph Bock for providing *C. reinhardtii* UVM11 and pJR39, Dr. Ian Blaby for critical reading of the manuscript, and Britany Reddish for technical support during protocol development.

References

1. Shimomura O, Johnson FH, Saiga Y (1962) Extraction, purification and properties of aequorin, a bioluminescent protein from the luminous hydromedusan, Aequorea. J Cell Comp Physiol 59:223–239
2. Prasher DC, Eckenrode VK, Ward WW et al (1992) Primary structure of the *Aequorea victoria* green-fluorescent protein. Gene 111:229–233
3. Chalfie M, Tu Y, Euskirchen G et al (1994) Green fluorescent protein as a marker for gene expression. Science 263:802–805
4. Gage DJ, Bobo T, Long SR (1996) Use of green fluorescent protein to visualize the early events of symbiosis between *Rhizobium meliloti* and alfalfa (*Medicago sativa*). J Bacteriol 178:7159–7166
5. Chiu W, Niwa Y, Zeng W et al (1996) Engineered GFP as a vital reporter in plants. Curr Biol 6:325–330
6. Pang SZ, DeBoer DL, Wan Y et al (1996) An improved green fluorescent protein gene as a vital marker in plants. Plant Physiol 112:893–900
7. Kain SR, Adams M, Kondepudi A et al (1995) Green fluorescent protein as a reporter of gene expression and protein localization. BioTechniques 19:650–655
8. Tsien RY (1998) The green fluorescent protein. Annu Rev Biochem 67:509–544
9. Ormö M, Cubitt AB, Kallio K et al (1996) Crystal structure of the *Aequorea victoria* green fluorescent protein. Science 273:1392–1395
10. Nagai T, Ibata K, Park ES et al (2002) A variant of yellow fluorescent protein with fast and efficient maturation for cell-biological applications. Nat Biotechnol 20:87–90
11. Neupert J, Karcher D, Bock R (2009) Generation of *Chlamydomonas* strains that efficiently express nuclear transgenes. Plant J 57:1140–1150
12. Rasala BA, Barrera DJ, Ng J et al (2013) Expanding the spectral palette of fluorescent proteins for the green microalga *Chlamydomonas reinhardtii*. Plant J 74:545–556
13. Barahimipour R, Strenkert D, Neupert J (2015) Dissecting the contributions of GC content and codon usage to gene expression in the model alga *Chlamydomonas reinhardtii*. Plant J 84(4):704–717
14. Lauersen KJ, Kruse O, Mussgnug JH (2015) Targeted expression of nuclear transgenes in *Chlamydomonas reinhardtii* with a versatile, modular vector toolkit. Appl Microbiol Biotechnol 99:3491–3503
15. Debuchy R, Purton S, Rochaix JD (1989) The argininosuccinate lyase gene of *Chlamydomonas reinhardtii*: an important tool for nuclear transformation and for correlating the genetic and molecular maps of the ARG7 locus. EMBO J 8:2803–2809
16. Kindle KL, Schnell RA, Fernández E, Lefebvre PA (1989) Stable nuclear transformation of Chlamydomonas using the Chlamydomonas gene for nitrate reductase. J Cell Biol 109:2589–2601
17. Cerutti H, Johnson AM, Gillham NW, Boynton JE (1997) Epigenetic silencing of a foreign gene in nuclear transformants of Chlamydomonas. Plant Cell 9:925–945
18. Hanikenne M, Merchant SS, Hamel P (2009) Transition metal nutrition: a balance between deficiency and toxicity. In: Stern D (ed) The *Chlamydomonas* sourcebook, vol 2, 2nd edn. Academic Press, San Diego, CA, pp 333–399
19. Blaby-Haas C, Merchant S (2013) Metal homeostasis: sparing and salvaging metals in chloroplasts. In: Culotta V (ed) Metals in cells. Encyclopedia of inorganic and

bioinorganic chemistry. John Wiley & Sons Ltd., Chichester, West Sussex, pp 51–64

20. Blaby-Haas CE, Merchant SS (2013) Iron sparing and recycling in a compartmentalized cell. Curr Opin Microbiol 16(6):677–685
21. Blaby-Haas CE, Merchant SS (2012) The ins and outs of algal metal transport. BBA-Mol Cell Res 1823:1531–1552
22. Merchant S, Allen M, Kropat J et al (2006) Between a rock and a hard place: trace element nutrition in Chlamydomonas. BBA-Mol Cell Res 1763:578–594
23. Malasarn D, Kropat J, Hsieh SI et al (2013) Zinc deficiency impacts CO_2 assimilation and disrupts copper homeostasis in *Chlamydomonas reinhardtii.* J Biol Chem 288:10672–10683
24. Urzica EI, Casero D, Yamasaki H et al (2012) Systems and trans-system level analysis identifies conserved iron deficiency responses in the plant lineage. Plant Cell 24:3921–3948
25. Castruita M, Casero D, Karpowicz SJ et al (2011) Systems biology approach in Chlamydomonas reveals connections between copper nutrition and multiple metabolic steps. Plant Cell 23:1273–1292
26. Deng X, Eriksson M (2007) Two iron-responsive promoter elements control expression of FOX1 in *Chlamydomonas reinhardtii.* Eukaryot Cell 6:2163–2167
27. Quinn JM, Merchant S (1995) Two copper-responsive elements associated with the Chlamydomonas Cyc6 gene function as targets for transcriptional activators. Plant Cell 7:623–628
28. Quinn JM, Merchant S (1998) Copper-responsive gene expression during adaptation to copper deficiency. Methods Enzymol 297:263–279
29. Gorman DS, Levine RP (1965) Cytochrome f and plastocyanin: their sequence in the photosynthetic electron transport chain of *Chlamydomonas reinhardi.* Proc Natl Acad Sci U S A 54:1665–1669
30. Kropat J, Hong-Hermesdorf A, Casero D et al (2011) A revised mineral nutrient supplement increases biomass and growth rate in *Chlamydomonas reinhardtii.* Plant J 66:770–780
31. Neupert J, Shao N, Lu Y, Bock R (2012) Genetic transformation of the model green alga *Chlamydomonas reinhardtii.* Methods Mol Biol 847:35–47
32. Kindle KL (1990) High-frequency nuclear transformation of *Chlamydomonas reinhardtii.* Proc Natl Acad Sci U S A 87:1228–1232
33. Harris E (1989) The chlamydomonas sourcebook. A comprehensive guide to biology and laboratory use. Academic Press, San Diego, CA
34. Blainey P, Krzywinski M, Altman N (2014) Points of significance: replication. Nat Methods 11:879–880
35. Glaesener AG, Merchant SS, Blaby-Haas CE (2013) Iron economy in *Chlamydomonas reinhardtii.* Front Plant Sci 4:337
36. Stramski D, Kiefer D (1991) Light-scattering by microorganisms in the open ocean. Prog Oceanogr 28:343–383
37. de Ruijter N, Verhees J, van Leeuwen W, van der Krol A (2003) Evaluation and comparison of the GUS, LUC and GFP reporter system for gene expression studies in plants. Plant Biol 5:103–115
38. Fuhrmann M, Hausherr A, Ferbitz L et al (2004) Monitoring dynamic expression of nuclear genes in *Chlamydomonas reinhardtii* by using a synthetic luciferase reporter gene. Plant Mol Biol 55:869–881
39. Shao N, Bock R (2008) A codon-optimized luciferase from *Gaussia princeps* facilitates the *in vivo* monitoring of gene expression in the model alga *Chlamydomonas reinhardtii.* Curr Genet 53:381–388
40. Ruecker O, Zillner K, Groebner-Ferreira R, Heitzer M (2008) Gaussia-luciferase as a sensitive reporter gene for monitoring promoter activity in the nucleus of the green alga *Chlamydomonas reinhardtii.* Mol Gen Genomics 280:153–162
41. Baek K et al (2016) DNA-free two-gene knockout in *Chlamydomonas reinhardtii* via CRISPR-Cas9 ribonucleoproteins. Sci Rep 6:30620
42. Shin SE et al (2016) CRISPR/Cas9-induced knockout and knock-in mutations in *Chlamydomonas reinhardtii.* Sci Rep 6:27810

Chapter 11

Gene Expression Analysis by Arylsulfatase Assays in the Green Alga *Chlamydomonas reinhardtii*

Ian K. Blaby and Crysten E. Blaby-Haas

Abstract

Chlamydomonas reinhardtii, a single-celled green alga, is a powerful microbial experimental system for understanding gene function. As a consequence of a high-quality genome sequence, community-wide efforts for gene model refinement and annotation, resources for strain collections and robust molecular techniques, research with this organism has significantly expanded in the past few decades. In two companion chapters, we outline colorimetric and fluorescence-based methodologies for genetic reporter systems in Chlamydomonas, which can be used to investigate and delineate gene expression and regulatory mechanisms. Here, we describe protocols for arylsulfatase activity assays using *ARS2*, activity of which can be measured either quantitatively or qualitatively, and in low (individual sample) or high (96-well format) throughput.

Key words Algae, *ARS2*, Gene fusion, Reporter, High-throughput assay

1 Introduction

Chlamydomonas reinhardtii (Chlamydomonas) is a unicellular alga of the Chlorophyte lineage that can be found in either the soil or fresh water. The organism was initially isolated over half a century ago, and a rich body of literature is available on all aspects of its biology [1]. As a consequence of its unique phylogenetic position among well-characterized organisms [2], and aided by a comprehensive repertoire of molecular and genetic techniques [3, 4], Chlamydomonas is often an organism of choice for addressing fundamental biological questions as they relate to both the plant and animal lineages (for example, photosynthesis and cilia, respectively). In recent years many studies have exploited the advanced genomics capabilities of the organism, enabling questions to be addressed at the -omics scale. Numerous illuminating systems-level and trans-omic studies have resulted from these approaches (recent examples: [5–7]). However, the cost (both in terms of time and finances) associated with these large-scale studies can limit their

Robert Damoiseaux and Samuel Hasson (eds.), *Reporter Gene Assays: Methods and Protocols*, Methods in Molecular Biology, vol. 1755, https://doi.org/10.1007/978-1-4939-7724-6_11, © Springer Science+Business Media, LLC, part of Springer Nature 2018

ability to rapidly assess individual genes under many conditions where genome-scale data is not required. Another factor to consider is that most cDNA sequencing (i.e., RNA-Seq) techniques do not discriminate between transcriptional and post-transcriptional regulation. Furthermore, mutational analyses, in which the upstream regions of genes can be scrutinized at the nucleotide level, are more suited to experiments on the individual gene scale.

Gene reporter systems can be used to interrogate genetic regulatory elements as well as discriminating conditions of gene expression. Moreover, if one assumes that the signal output is proportional to the transcript level of the target gene, the extent of gene expression can be extrapolated. Several reporter systems have been developed in Chlamydomonas that can be used for expression studies, employing fluorescence [8, 9], luminescence [10, 11] and colorimetric outputs. Arylsulfatase assays, first described in Chlamydomonas by de Hostos et al., rely on genetic fusions between the native Chlamydomonas *ARS2* gene and the upstream region of a gene of interest [12]. *ARS2* encodes a periplasmic arylsulfatase, which in its native genomic context is induced only under conditions of sulfur deprivation [13], and enables the scavenging of sulfur from the natural environment by catalyzing the cleavage of sulfate groups from aromatic substrates. The activity of ARS2 can be detected in the presence of a chromogenic substrate [14], such as *para*-nitrophenylsulfate, 5-bromo-4-chloro-3-indolyl sulfate (X-SO_4) and α-naphthylsulfate [15]. By fusing the promoterless *ARS2* gene to a promoter of interest, both quantitative and qualitative assays can be performed using cleavage of the chromogenic substrate as a proxy for promoter activity [14, 15]. A major advantage of these colorimetric methods over the alternative fluorescence or luciferase-based systems is that, at least for qualitative assays, no specialist apparatus such as a plate reader or spectrophotometer is required. The unicellular nature of Chlamydomonas lends itself to experimental scale-up using 96- and 384-well plate formats. This has the benefits of increasing throughput of experimental conditions and replicate number whilst simultaneously reducing individual sample processing/analysis time, costs of generating large volumes of growth medium, and the required incubator capacity. Additionally, analyzing dozens of independent transformants of a single construct provides robust statistical power to overcome inherent noise in expression level resulting from genome insertion positional effects. Herein, we outline considerations, experimental designs and protocols for the deployment of this organism and utilization of gene expression studies using arylsulfatase assays. Readers are referred to the previous chapter for protocols utilizing fluorescent proteins in Chlamydomonas reporter assays.

2 Materials, Equipment, and Experimental Considerations

1. *C. reinhardtii* str. CC-425 (genotype: *cw15 arg2 sr-u-2-60 mt*$^+$); catalogued strains of Chlamydomonas can be obtained from the Chlamydomonas resource center (http://chlamycollection.org)
2. Growth conditions:

   Due to the slow growth rate of Chlamydomonas relative to other microorganisms, it is strongly advised that transfer and inoculation of media with Chlamydomonas be performed under sterile conditions (ideally in a laminar flow hood, but minimally in the close presence of a Bunsen burner flame). Aseptic technique and clean, organized working practices should be applied at all stages. All Chlamydomonas liquid cultures are grown at 23 °C, ~80 μmol m^{-2} s^{-1}, with rotational shaking at 180 RPM, e.g., in Innova incubators (Eppendorf), or similar. Light is supplied by fluorescent tubes in a 1:2 ratio of warm (3000 K) to cool (4100 K). To ensure adequate aeration, flasks should contain no more than 20% medium of total flask volume. Static growth, either on petri dishes, Omni-trays or 96-well plates are grown at 23 °C, ~80 μmol m^{-2} s^{-1} in growth chambers (Percival CU-36L6 or similar). Prior to selection of transformation of vectors encoding ARG7, growth medium for str. CC-425 should be supplemented with 100 μg/mL arginine (filter sterilized if added subsequent to autoclaving, and 200 μg/mL arginine should be used if arginine is added prior to autoclaving).
3. Chlamydomonas growth medium:

   Tris-acetate phosphate (TAP) medium [16]: each liter contains 10 mL TAP salts solution (15 g NH_4Cl, 4 g $MgSO_4{\cdot}7H_2O$, 2 g $CaCl_2{\cdot}2H_2O$, final volume of 1 L H_2O), 8.3 mL phosphate solution (14.34 g K_2HPO_4, 7.26 g KH_2PO_4, final volume of 1 L H_2O), 10 mL Tris-acetate solution (242 g Tris, 100 mL glacial acetic acid, final volume of 1 L H_2O) and 1 mL each component of Kropat trace metal solutions [17].
4. Select agar (Thermo Fisher 30391-023); washed three times in purified H_2O. Use 15 g washed agar per 1 L of TAP.
5. Plasmids:

   pJD100GW (a Gateway-adapted version of pJD100 [18]; note that pJD100 was referred to as ptubB2Δ3,2,1/ars in the original publication [19]); pJD100 (and pJD100GW) contains a minimal β-tubulin promoter directly upstream of *ARS2*. pARG7.8 (for coselection of pJD100GW, *see* below). Both vectors are available from the Chlamydomonas resource center. pDONR221 is available from ThermoScientific, cat #12536-017.

6. *Escherichia coli* transformants are selected on agar-solidified LB medium containing 50 μg/mL kanamycin (for pDONR221 derivatives) or 100 μg/mL ampicillin (for pJD100GW derivatives).
7. TAPS medium:

   TAP medium containing sorbitol at a final concentration of 50 μM. Autoclave in 50 mL aliquots.
8. Plastics:

   Sterile, lidded, 96-well plates (eg Nunc cat# 267313 for colorimetric assays, transparent plates are adequate). Omni-trays (ThermoScientific, cat# 140156). Deep dish petri dishes (ThermoScientific, cat# 4031). 50 mL Falcon tubes are conveniently sized for culture centrifugation prior to transforming.
9. 2 mm gap electroporation cuvettes (e.g., FisherScientific cat# FB102).
10. Electroporation System (such as a Bio-Rad Gene Pulser).
11. LR clonase II and BP clonase II reaction kits (Life Technologies, cat# 12538-120 and 11789-020).
12. Small spray bottle (e.g., FisherScientific, cat# S06755).
13. Colorimetric reaction solution A:

    10 mM 5-Bromo-4-chloro-3-indolyl sulfate potassium salt (X-SO_4; Sigma, cat# B1379) in 0.1 M Tris–HCI, pH 7.5.
14. Colorimetric reaction solution B:

    20 mM alpha-Naphthyl sulfate potassium salt (MP Biomedicals, cat# 0210025801).
15. 4% SDS (w/v) in 0.2 M Na acetate buffer, pH 4.8
16. Colorimetric reaction solution C:

    0.1 M glycine–NaOH (pH 9.0), 10 mM imidazole, 4.5 mM p-nitrophenyl sulfate.
17. 2.5 M NaOH.
18. All protocol steps using 96-well plates can be significantly expedited by using an 8- or 12-channel pipette (e.g., Eppendorf 10–100 μL 12-channel pipette). Additionally, the reduced number of liquid transfers minimizes risk of missed wells and pipetting errors.
19. Plate reader capable of absorbance measurements at 750 nm and 540 nm, e.g., Tecan M1000 PRO.
20. Bench-top centrifuge with tube and 96-well plate format capabilities, e.g., Eppendorf 5810.

3 Methods

3.1 Construction of Reporter (See Fig. 1)

This protocol describes generation of promoter-fusion constructs in pJD100GW using Gateway® Technology. In general, a 500 bp fragment downstream of the gene (+500 to +1; where transcription start site is position 0) is sufficient to capture the promoter. Use the most accurate and up-to-date version of the Chlamydomonas genome sequence and gene models, presently v5 and v5.5 respectively, available on Phytozome (http://phytozome.jgi.doe.gov/pz/portal.html) to determine the transcription start site [20]. The accuracy of the gene model can be confirmed by comparison to publicly available expressed sequence tags (ESTs), which can be viewed in the genome browser hosted on Phytozome. To clone the promoter fragment into pJD100GW, amplify purified genomic DNA by PCR. The forward primer needs to contain four guanines (G) followed by the 25-nucleotide *att*B1 sequence (5′ ACAAGTTTGTACAAAAAAGCAGGCT 3′) and 18- to 25-nucleotide target-specific sequence. The reverse primer needs to contain four guanines followed by the 25-nucleotide *att*B2 sequence (5′ ACCACTTTGTACAAGAAAGCTGGGT 3′) and 18- to 25-nucleotide target-specific sequence.

We have found Primer3Plus (http://www.bioinformatics.nl/cgi-bin/primer3plus/primer3plus.cgi/ [21]) to be a convenient and robust tool for selecting the target-specific oligonucleotide

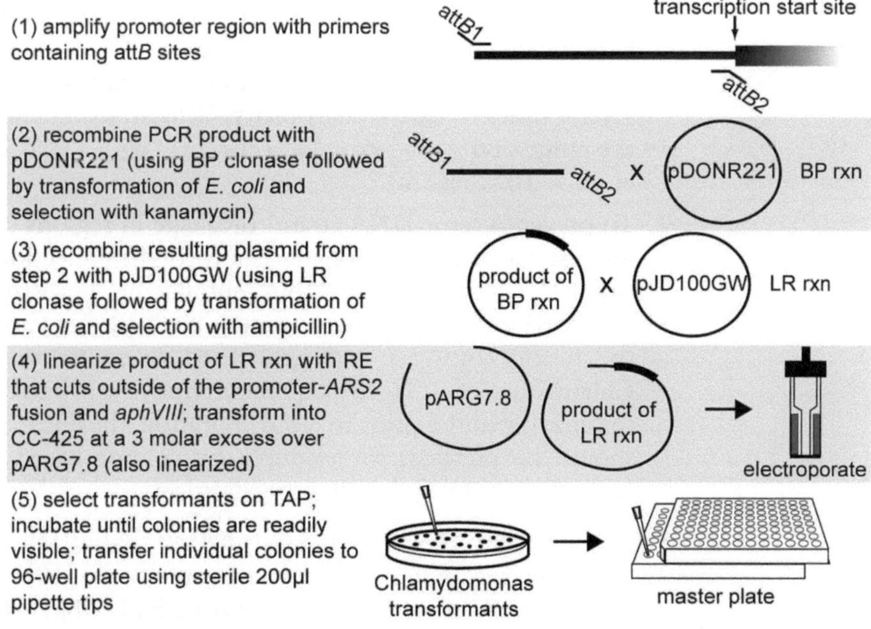

Fig. 1 Workflow and schematic of protocol steps 3.1–3.1.2. Panels 1–3 refer to protocol 3.1. Panel 4 refers to Subheading 3.1.1 and panel 5 refers to Subheading 3.1.2. Reaction is abbreviated to rxn

sequences, with the following changes to default parameters for amplification from the Chlamydomonas genome: primer size Min: 21, Opt: 24, Max: 27 and primer Tm Min 59, Opt: 62, Max: 65. We also routinely use Phusion High-Fidelity DNA Polymerase with Phusion GC buffer (ThermoScientific cat# F-530S) according to the manufacture's protocol, and add DMSO to a final concentration of 3% (v/v). The PCR product is then cloned using Gateway® Technology with pDONR™221 as the entry vector for the BP reaction (ThermoFisher Scientific cat# 12535-029) and pJD100GW as the destination vector in the LR reaction (ThermoFisher Scientific cat# 11791-020). Alternatively, traditional cloning techniques can be employed to insert the region of interest into the KpnI site of pJD100. Once the region of interest has been successfully cloned, regulatory sites can be identified and/or validated by introducing nucleotide mutations by a variety of techniques (*see* [22] for an efficient method).

Recent advances in chemical gene synthesis offer new capabilities in terms of throughput, experimental design and time/cost savings. DNA promoter fragments can be synthesized containing predesigned point mutations or deletions, and flanked by *att*B sites for direct cloning by Gateway® methods (without the need for a PCR step). In particular, G-blocks (IDT, IA) are convenient, rapidly synthesized, and cost-effective for this purpose.

3.1.1 Transformation by Electroporation

Expected efficiency: ~1 colony (i.e., a single transformant) per ng linearized DNA, *see* **Note 1**.

1. Inoculate str. CC-425 into TAP medium plus 100 μg/mL arginine. If inoculation is directly from a slant or plate that has been grown under day/night cycle conditions, subculture several times to lose synchronicity. Cells are competent within the range of 6×10^5–6×10^6 cells/mL, with a peak efficiency at ~8×10^5 cells/mL.
2. To prevent vector DNA strand breakage in regions of specific interest (such as the cloned region, or the selection marker (*aphVIII*)), supercoiled DNA should be linearized using appropriate endonuclease enzyme(s), which cleave outside of the fusion construct and gene used to confer selection in Chlamydomonas. To subsequently remove salts introduced in the enzyme buffer prior to electroporation, the linearized DNA should be purified, for example by column-based methods (e.g., Qiagen Qiaquick PCR purification).
3. Prechill electroporation cuvettes (gap size 2 mm) on ice.
4. Once the culture has achieved a density of ~8×10^5 cells/mL, transfer to 50 mL sterile tubes. Add 12.5 μL 20% (v/v) Tween 20 (dilution of 1:4000). Gently mix by inverting five times then incubate statically for 5 min.

5. Collect the cells by centrifugation 1900 × *g*, 5 min, room temperature. Pour off the supernatant.
6. Gently resuspend the pellet in 10 mL TAPS by slow pipette action.
7. Collect the cells by centrifugation 1900 × *g*, 5 min, room temperature. Pour off the supernatant.
8. Gently resuspend the pellet by slow pipette action in TAPS to a final cell count of 4×10^8 cells/mL.
9. Add 300 ng of linearized pARG7.8 to 3 separate electroporation cuvettes. Label the three cuvettes as negative control, positive control and test. Then, add a threefold molar excess of linearized (1) negative control, such as pJD100GW; (2) positive control (pJD100GW containing a promoter fragment that leads to inducible *ARS2* expression during growth in the test condition) and (3) test construct to each cuvette, respectively. Use additional cuvettes as required for the number of test constructs.
10. Transfer 150 μL concentrated Chlamydomonas cells to each electroporation cuvette. Gently tap to mix cells and DNA. Place on ice for 5 min.
11. Wipe off any melted ice from cuvette, and electroporate using the settings: 800 V, 1000 Ω, 25 μF (settings optimized for a BioRad Gene Pulser). As quickly as possible add 100 μL TAPS. Having a pipette tip preloaded can save a few seconds. Time constants (as collected in a Bio-Rad Gene Pulser) should be in the range of 9–11.
12. Incubate the cells in the electroporation cuvettes for 30 min room temperature, static.
13. Gently collect the cells using a pipette, transfer to a 15 mL sterile tube preloaded with 5 mL TAPS. Place tubes on a benchtop tube rotator and agitate for 8 h, room temp, 20–-40 μmol m^{-2} s^{-1} (a lower light intensity is advised for this step since division of cells will result in nonindependent transformants, and is undesirable for assay statistical robustness).
14. Collect the cells by centrifugation 300 × *g*, 5 min, room temperature. Pour off the supernatant and gently resuspend the pellet in the remaining ~200 μL TAPS. Pipette cells onto TAP agar (without arginine) and gently spread using a sterilized (e.g., flamed, and subsequently cooled) loop.
15. Incubate plates in growth chamber. Colonies should be visible in 7–12 days.

3.1.2 Rearray Transformants

Transformed DNA near-randomly inserts primarily by illegitimate recombination into the Chlamydomonas genome, resulting in position-specific effects on the expression level of the reporter construct. Consequently, each transformant will have different levels of arylsulfatase activity, dependent on the position of insertion. Additionally, because pJD100GW lacks a selection marker for Chlamydomonas, transformants are selected by cotransformation with a marker (pARG7.8 in protocols above). As a result, not every arginine prototroph may have been successfully transformed with the reporter. To overcome these limitations, we routinely screen 96 independent transformants per construct, by inoculating a 96-well plate, which can then be used as a "master plate" (Fig. 1), which is a starting point for assay steps 3.2.1 and 3.2.2. For the same reasons, 96 independent transformants of the pJD100GW negative control should also checked for cotransformation efficiency, and should be carried forward through subsequent steps.

1. Under aseptic conditions, fill a standard 96-well plate with 200 μL TAP per well.
2. Inoculate each well with an independent transformant colony. Autoclaved pipette tips are useful for this purpose. Ensure transfer of cells by swirling tips in the well.
3. Incubate plates in growth chamber. Inoculated wells should have grown sufficiently to be visibly green in ~7 days (forming the master plate for steps 3.2.1 and 3.2.2).
4. To determine whether the reporter construct was successfully cotransformed with pARG7.8, transfer 10 μL cell culture from each well to a new 96-well plate. Confirm presence of pJD100GW by PCR using a primer pair designed to anneal uniquely to pJD100GW and amplify a product of roughly 250–500 bp *see* **Note 2**.

3.2 Arylsulfatase Assay (See Fig. 2)

3.2.1 Qualitative (i.e., End-Point) Assay

This assay is used to quickly screen independent promoter constructs and identify regions of the promoter fragment that are required for condition-specific expression. Because of the site-of-insertion position effects (*see* above, and previous chapter for further discussion), this assay can also be used to screen transformants for expression level, and accordingly identify clones to carry forward for low-throughput kinetic assay (3.2.3 below), or to quickly screen constructs on an express/does not express basis.

1. Fill two Omni-trays (55 mL per tray), one with regular TAP agar and the second with agar-solidified medium that will lead to induction of your promoter.
2. Ensure cells in master plate, growing in liquid media, are uniformly distributed in the well. If they are not, gently resuspend by pipetting with sterile tips.
3. Using a multichannel pipette, transfer 5 μL of cells from the 96-well master plate to both Omni-trays. Take care to ensure the drops do not flood into each other, and that the agar surface is dry before moving the plate.

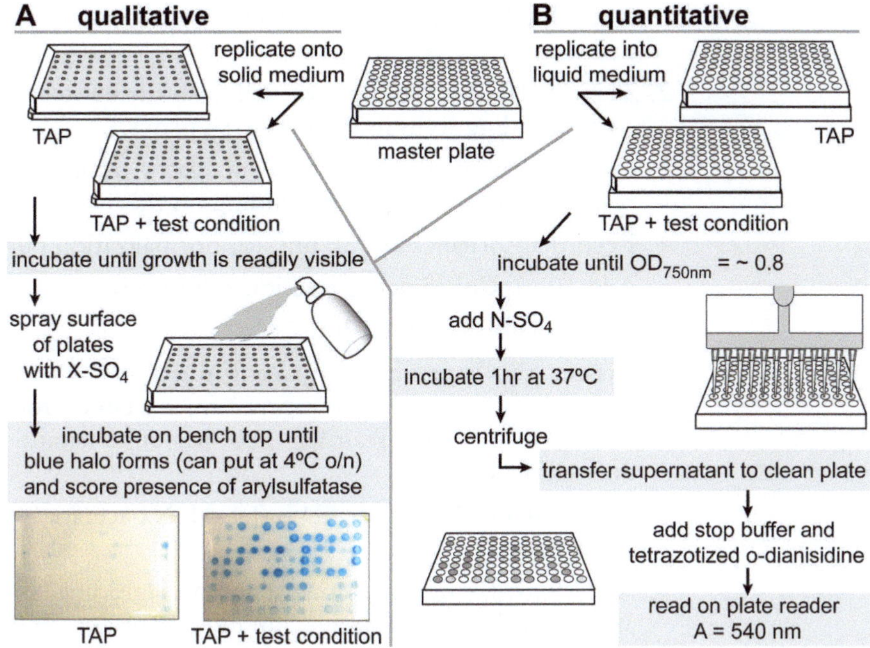

Fig. 2 Workflow and schematic of protocols 3.2.1 and 3.2.2. Starting with a master plate containing 96 individual *C. reinhardtii* arginine prototrophs, transformants can be transferred with a multichannel pipette to agar-solidified plates containing control and test media to perform a qualitative assay (**a**) or transferred to 96-well plates containing control and test liquid media to perform a quantitative assay (**b**). Bottom left, an example of the blue halos visualized after incubation with the substrate; cells were removed from the surface of an Omnitray filled with agar-solidified uninducing (TAP) or inducing (TAP + test condition) medium; each position (1–96) represents an independent transformant

4. Incubate the two Omni-trays in growth chamber for 7 days.
5. Uniformly spray surface of the agar with Colorimetric solution A until all colonies are lightly covered. Allow the plates to sit on a benchtop for 6–24 h. Cells that are expressing the arylsulfatase will have a blue halo around them in the agar *see* **Note 3**.

3.2.2 Quantitative Assay (End-Point)—High-Throughput

In addition to providing data acquired by the qualitative assay above, this assay is also used to screen for promoter constructs that no longer regulate arylsulfatase activity under the test condition, to quantify the difference between different constructs, or to pick transformants for the low-throughput kinetic assay (3.2.3 below). The largest difference between the two assays is that the level of arylsulfatase activity can be quantified. This assay can also be used to screen activity of a promoter fragment under various growth conditions or stresses.

For each construct or master plate perform the following steps:

1. Using a multichannel pipette and basin, fill one sterile 96-well plate with 200 μL TAP per well and a second 96-well plate with

liquid medium that will lead to absence of repression of your promoter.

2. Transfer 10 μL of culture per well from the master plate to the corresponding wells in the two fresh plates.
3. Incubate the plates in a static incubator until the OD_{750nm} reaches ~0.8 (~3–6 days, depending on the inoculum density).
4. Record absorbance at 750 nm, for normalization purposes
5. Add 10 μL Colorimetric reaction solution B.
6. Incubate at 37 °C for 1 h.
7. Collect the supernatant by centrifugation of the plates at 200 × *g* for 2 min, room temperature, in a bench-top swinging bucket centrifuge. Transfer 100 μL supernatant to clean 96-well plate.
8. Add 100 μL 4% (w/v) SDS/0.4 M Na-acetate, pH 4.8 to supernatant.
9. Add 20 μL 10 mg/mL tetrazotized o-dianisidine and immediately read absorbance at 540 nm in plate reader. Save data.

3.2.3 Quantitative Assay (Time Course)—Low-Throughput (See Fig. 3)

1. Inoculate three independent transformants from the master plate into TAP-containing flasks to a density of ~ 1×10^5 cells/mL.
2. Once the cultures reach 1×10^6 cells/mL, transfer half of each culture into 50 mL Falcon tubes. Collect cells by centrifugation (1900 × *g*, 5 min, room temp). Resuspend half the culture in control medium and half the culture in test medium (6 cultures total: 3 biological replicates in control medium and 3 biological replicates in test medium).
3. Incubate culture flasks in shaking incubator for 24 h.
4. Transfer 50 μL of cell culture to a new 15 mL tube and add 500 μL Colorimetric reaction solution C. Repeat for each of the six cultures. At this stage, also take a sample from each culture to determine cell density (cells/mL) for normalization of arylsulfatase activity between samples.
5. Incubate tube for 30 min at 27 °C.
6. Quench the reaction by addition 2 mL of 2.5 M NaOH to each tube 15 mL tube.
7. Pellet the cells by centrifugation of the reaction mixture at 1900 × *g* for 5 min, room temp.
8. Transfer supernatant to spectrophotometer cuvette, and measure the absorbance at 410 nm.
9. Repeat **steps 4–8** after 2, 4, 8, 12, 24, 48, and 72 h. Note that these times may be gene-dependent, and may require some optimization. Save all data.

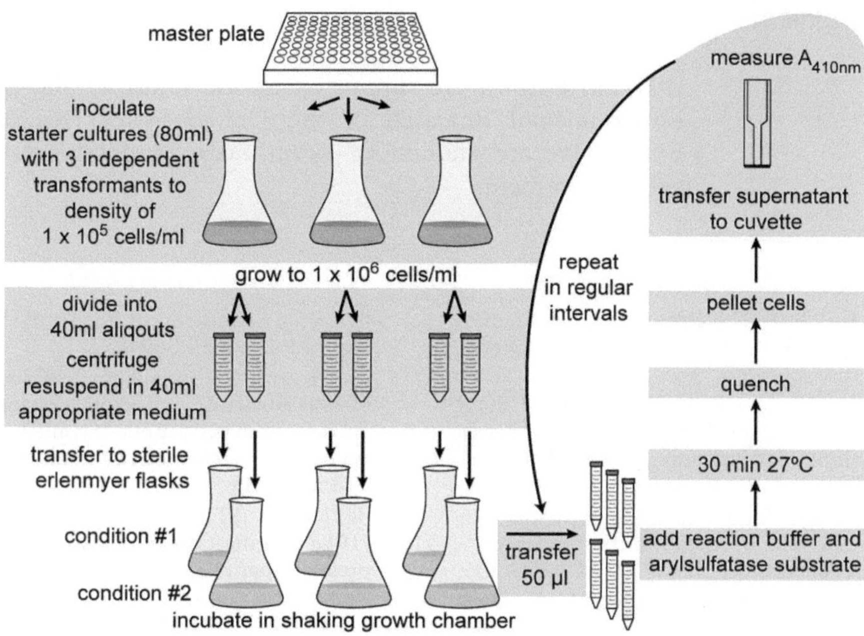

Fig. 3 Workflow and schematic of protocol step 3.2.3. Starting with a master plate containing 96 individual *C. reinhardtii* arginine prototrophs, three independent transformants, whose condition-specific arylsulfatase activity was confirmed as in Fig. 2, are used to start three separate cultures. Cells from these three cultures are subsequently grown in test and control media. Samples are collected at regular intervals, and arylsulfatase activity is measured in a low-throughput quantitative assay

10. Normalize OD_{410} values to cell density. Generate a plot of normalized absorbance by time.

4 Notes

1. Researchers should also note the excellent protocols for Chlamydomonas nucleus genome transformations by Neupert et al. [23].
2. Colony PCR methods, as described by Cao et al., avoid the need to prepare genomic DNA, and can be easily scaled to 96-well format [24]. We have found that approaching 100% of transformants contain both plasmids. Should the researcher find that the cotransformation efficiency is lower, he/she may wish to array additional transformants, validated to contain the reporter construct, to maximize the plate capacity in downstream assays.
3. If the blue halo is difficult to see after 24 h, the cells can be removed from the agar by gently applying H_2O (e.g., from either a serological pipette or squeeze-bottle) or gently wiping away cells with a Kimwipe™.

Acknowledgments

IKB and CEB-H are supported by the Office of Biological and Environmental Research of the United States Department of Energy. We are grateful to Kevin Keck for critical reading and helpful comments.

References

1. Harris E (2009) The chlamydomonas sourcebook, 2nd edn. Academic Press, San Diego, CA
2. Merchant SS, Prochnik SE, Vallon O et al (2007) The Chlamydomonas genome reveals the evolution of key animal and plant functions. Science 318:245–250
3. Jinkerson RE, Jonikas MC (2015) Molecular techniques to interrogate and edit the Chlamydomonas nuclear genome. Plant J 82:393–412
4. Mussgnug JH (2015) Genetic tools and techniques for *Chlamydomonas reinhardtii*. Appl Microbiol Biotechnol 99:5407–5418
5. Zones J, Blaby I, Merchant S, Umen J (2015) High-resolution profiling of a synchronized diurnal transcriptome from *Chlamydomonas reinhardtii* reveals continuous cell and metabolic differentiation. Plant Cell 27:2743–2769
6. Mettler T, Mühlhaus T, Hemme D et al (2014) Systems analysis of the response of photosynthesis, metabolism, and growth to an increase in irradiance in the photosynthetic model organism *Chlamydomonas reinhardtii*. Plant Cell 26:2310–2350
7. Ngan CY, Wong C-H, Choi C et al (2015) Lineage-specific chromatin signatures reveal a regulator of lipid metabolism in microalgae. Nat Plants 1(8). https://doi.org/10.1038/nplants.2015.107
8. Fuhrmann M, Oertel W, Hegemann P (1999) A synthetic gene coding for the green fluorescent protein (GFP) is a versatile reporter in *Chlamydomonas reinhardtii*. Plant J 19:353–361
9. Rasala BA, Barrera DJ, Ng J et al (2013) Expanding the spectral palette of fluorescent proteins for the green microalga *Chlamydomonas reinhardtii*. Plant J 74:545–556
10. Shao N, Bock R (2008) A codon-optimized luciferase from *Gaussia princeps* facilitates the in vivo monitoring of gene expression in the model alga *Chlamydomonas reinhardtii*. Curr Genet 53:381–388
11. Fuhrmann M, Hausherr A, Ferbitz L et al (2004) Monitoring dynamic expression of nuclear genes in *Chlamydomonas reinhardtii* by using a synthetic luciferase reporter gene. Plant Mol Biol 55:869–881
12. de Hostos EL, Schilling J, Grossman AR (1989) Structure and expression of the gene encoding the periplasmic arylsulfatase of *Chlamydomonas reinhardtii*. Mol Gen Genet 218:229–239
13. de Hostos EL, Togasaki RK, Grossman A (1988) Purification and biosynthesis of a derepressible periplasmic arylsulfatase from *Chlamydomonas reinhardtii*. J Cell Biol 106:29–37
14. Davies JP, Weeks DP, Grossman AR (1992) Expression of the arylsulfatase gene from the beta 2-tubulin promoter in *Chlamydomonas reinhardtii*. Nucleic Acids Res 20:2959–2965
15. Ohresser M, Matagne RF, Loppes R (1997) Expression of the arylsulphatase reporter gene under the control of the *nit1* promoter in *Chlamydomonas reinhardtii*. Curr Genet 31:264–271
16. Gorman DS, Levine RP (1965) Cytochrome *f* and plastocyanin: their sequence in the photosynthetic electron transport chain of *Chlamydomonas reinhardi*. Proc Natl Acad Sci U S A 54:1665–1669
17. Kropat J et al (2011) A revised mineral nutrient supplement increases biomass and growth rate in *Chlamydomonas reinhardtii*. Plant J 66:770–780
18. Allen MD, Kropat J, Merchant SS (2008) Regulation and localization of isoforms of the aerobic oxidative cyclase in *Chlamydomonas reinhardtii*. Photochem Photobiol 84:1336–1342
19. Davies JP, Yildiz F, Grossman AR (1994) Mutants of chlamydomonas with aberrant responses to sulfur deprivation. Plant Cell 6:53–63
20. Goodstein DM, Shu S, Howson R, Neupane R, Hayes RD, Fazo J, Mitros T, Dirks W, Hellsten U, Putnam N, Rokhsar DS (2012) Phytozome: a comparative platform for green plant genomics. Nucleic Acids Res 40:D1178–D1186
21. Rozen S, Skaletsky H (2000) Primer3 on the WWW for general users and for biologist programmers. Methods Mol Biol 132:365–386

22. Hemsley A, Arnheim N, Toney MD et al (1989) A simple method for site-directed mutagenesis using the polymerase chain reaction. Nucleic Acids Res 17:6545–6551
23. Neupert J, Shao N, Lu Y, Bock R (2012) Genetic transformation of the model green alga *Chlamydomonas reinhardtii*. Methods Mol Biol 847:35–47
24. Cao M, Fu Y, Guo Y, Pan J (2009) Chlamydomonas (Chlorophyceae) colony PCR. Protoplasma 235:107–110

Chapter 12

Endogenous Locus Reporter Assays

Yaping Liu, Jeffrey Hermes, Jing Li, and Matthew Tudor

Abstract

Reporter gene assays are widely used in high-throughput screening (HTS) to identify compounds that modulate gene expression. Traditionally a reporter gene assay is built by cloning an endogenous promoter sequence or synthetic response elements in the regulatory region of a reporter gene to monitor transcriptional activity of a specific biological process (exogenous reporter assay). In contrast, an endogenous locus reporter has a reporter gene inserted in the endogenous gene locus that allows the reporter gene to be expressed under the control of the same regulatory elements as the endogenous gene, thus more accurately reflecting the changes seen in the regulation of the actual gene. In this chapter, we introduce some of the considerations behind building a reporter gene assay for high-throughput compound screening and describe the methods we have utilized to establish 1536-well format endogenous locus reporter and exogenous reporter assays for the screening of compounds that modulate Myc pathway activity.

Key words Endogenous locus reporter, Exogenous reporter, High-throughput screening (HTS), Reporter gene, Myc, Luciferase, NanoLuc, PEST

1 Introduction

Over the last decade, the reporter gene assay has become a powerful method in high-throughput compound screening programs due to its high sensitivity, automation compatibility and low cost [1–3]. A reporter gene assay can be quickly developed and miniaturized to high density high-throughput screening (HTS) format for identifying compounds that modulate a specific target or biological process, such as a signaling pathway activity.

Traditionally a reporter gene assay uses exogenous plasmid based overexpression system in which the promoter sequence is derived from endogenous transcription factor binding sites or synthetic response elements [4–6]. The promoter sequence can be cloned from genomic DNA samples (e.g., DNA library screening, PCR amplification, etc.), or de novo synthesized, and then constructed in a reporter vector to drive the expression of a reporter gene, such as GFP, β-galactosidase, luciferase, and beta-lactamase [7]. The various regions of endogenous promoter sequence or

Robert Damoiseaux and Samuel Hasson (eds.), *Reporter Gene Assays: Methods and Protocols*, Methods in Molecular Biology, vol. 1755, https://doi.org/10.1007/978-1-4939-7724-6_12,

multiple copies of synthetic response element can be used to make reporter vectors that optimize reporter expression level and assay performance. Such reporter vectors can be transiently or stably transfected into cells to monitor reporter gene transcription [4]. It can also be used in different cell lines to assess gene expression and pathway regulation in different biological background. However this approach may result in artifactual results as it is limited to the proximal promoter sequence and it is not possible to include other important gene regulating elements such as distal enhancers and chromosomal structure [8]. Also, the choice of promoter sequence may over emphasize one regulating element while neglecting others, and thus not capture the true biology. Finally, transiently transfected plasmids are not subject to the same epigenetic regulation via chromatin modifications as chromosomal loci, and stably transfected randomly integrated reporters adopt the chromatin state of their host locus rather than that of the gene of interest.

Recent development in gene editing technologies, such as CRISPR [9, 10], TALEN [11, 12], and recombinant Adeno-Associated Virus (rAAV) [13–15], has made it possible to introduce a reporter gene into a specific site of a endogenous gene locus to generate stable cell lines that express the reporter gene. For example, Horizon Discovery (Cambridge, UK) has developed a series of endogenous locus reporter cell lines (X-MAN NanoLuc reporter kits), utilizing their proprietary rAAV gene editing technology that allows the precise editing of the genome of mammalian somatic cells down to single base-pair resolution [15, 16]. In an X-MAN NanoLuc reporter line, a novel NanoLuc luciferase (Promega, WI), reporter gene is knocked-in to a single allele of the endogenous gene locus, thus allowing the reporter gene expression to be under the control of the same regulatory elements as the endogenous gene. This approach will likely capture broader gene regulation and more accurately reflect the changes seen in real disease biology. However it may have the following drawbacks: (1) the expression level of reporter gene may be low due to weak endogenous promoter activity, resulting in lower assay signal-to-background ratio that may not be amendable to miniaturization and automation; (2) disruption of endogenous locus or off-target integration may causes cell viability defects [17, 18]; (3) it takes longer time to generate a reporter cell line and the associated cost is much higher than a traditional exogenous plasmid reporter assay; (4) the available cell lines for a given pathway reporter is often limited; (5) the method makes certain frequently used techniques for characterization of regulation such as sequential/nested deletion analysis (*see* **Note 1**) much more difficult, thus hampering analysis of causality.

Myc is a transcription factor that serves as a master regulator of cell proliferation, differentiation, and apoptosis [19, 20]. It was estimated that Myc binds to 15% of the genome and regulates the expression of its many target genes [21]. Myc is a potent oncogene that can promote tumorigenesis in a wide range of tissues. Elevated expression of Myc occurs frequently in human cancers and is associated with tumor aggressiveness and poor clinical outcome [22]. Myc expression is regulated by many signaling pathways at the transcriptional level, including Wnt/β-catenin, ERK/MAPK, and TGFβ/SMAD pathways [19], however the optimal nodes for targeting Myc pathway are not known. In this chapter we use Myc reporter gene assay as an example to introduce some of the considerations on building a reporter assay for high-throughput compound screening. We also describe the methods we have used to establish 1536-well format endogenous locus reporter and exogenous promoter reporter assays for the screening of compounds that modulate Myc pathway activity.

1.1 Design of Reporter Assay

1.1.1 Design of Exogenous Reporter Assay

To design a promoter reporter assay to be used to monitor transcriptional activity of a gene or signaling pathway, we first need to decide which promoter sequence or synthetic response element to use in the reporter vector for driving reporter gene expression. Myc proximal promoter sequence is estimated to be a 2.5 kb genomic fragment upstream of Myc transcription start site [23]. Since the endogenous Myc locus reporter developed by Horizon is only available in HCT116, a human colorectal carcinoma cell line, we planned to use HCT116 for developing exogenous transient reporter assay as well. HCT116 cells harbor a low copy-number Myc gain and a stable mutant β-catenin leading to constitutive Myc overexpression [23–25]. In HCT116 and other colorectal cancer cell lines, the two TCF binding elements (TBE) at 1.2 kb and 0.6 kb upstream of the TATA box at the Myc major transcription start site are believed to be responsible for APC/β-catenin pathway mediated Myc transcription activation [23], so it is preferable to use the Myc promoter fragment encompassing both TBEs as promoter to make exogenous reporter vectors (Fig. 1).

1.1.2 Ready-Made Reporter Vectors

The quickest way to acquire an exogenous reporter vector is through commercial sources or plasmid repository services. For example, SwitchGear Genomics (http://switchgeargenomics.com/) provides a genome-wide panel of transfection-ready LightSwitch reporter assays that includes over 18,000 endogenous promoters and synthetic response elements linked to a proprietary RenSP luciferase reporter gene. RenSP luciferase protein has a half-life of ~1 h, enabling a detailed analysis of kinetic responses with a highly robust signal.

A reporter vector of interest can be identified through online search of the LightSwitch promoter reporter GoClone collection. For example, when a gene name "Myc" is entered in the promoter searching field, one default and five alternative promoter reporter vectors are returned, each represents a different region of a

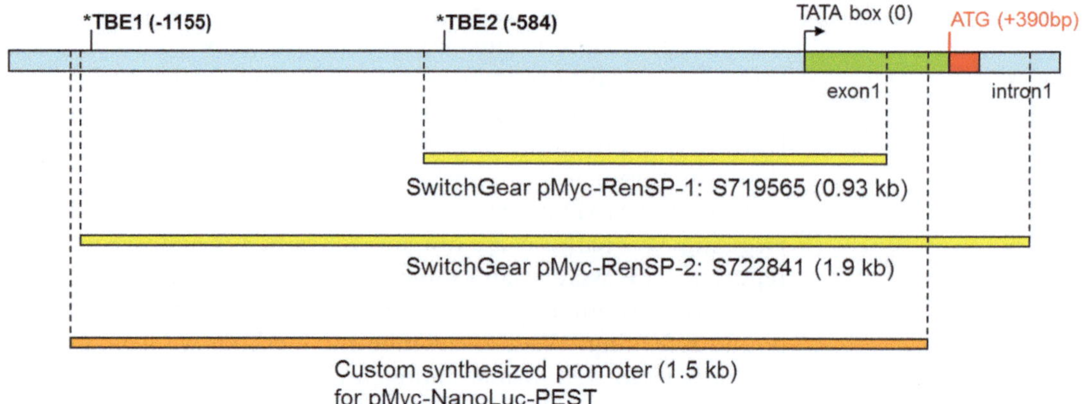

Fig. 1 Scheme of Myc gene promoter region and various reporter vector designs. Numbers in () indicate the distance (bp) upstream (−) or downstream (+) from Myc major transcription start site (TATA box)

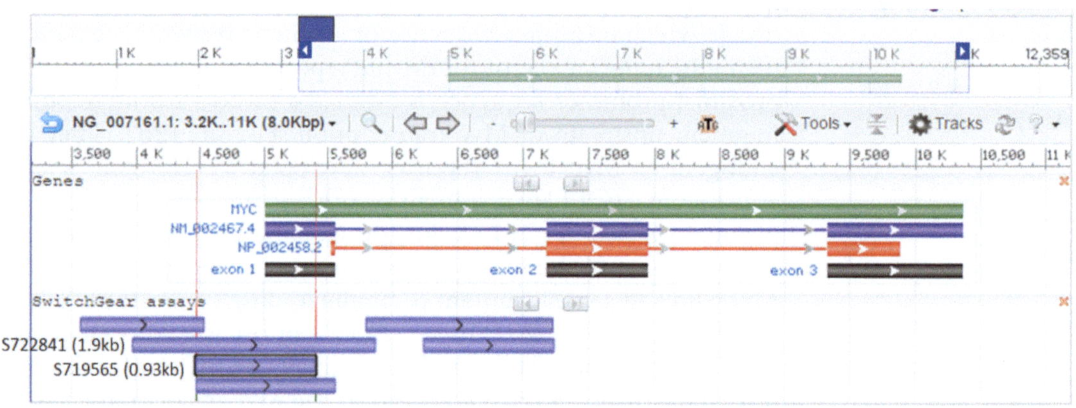

Fig. 2 DNA sequence alignment of LightSwitch promoter vector sequences to Myc gene sequence. The alignments of Myc genomic DNA, transcribed cDNA, protein-coding sequence, and exons are shown in the "Genes" section. Sequence alignments of the six SwitchGear reporter vectors to Myc gene are shown in the "SwitchGear assays" section, with S722841 (default vector) highlighted to illustrate its location relative to the transcription and translation start sites of Myc. This figure was made using NCBI's "Gene Graphics" view with "blastn" search results configured in the custom track (http:/www.ncbi.nlm.nih.gov/)

hypothetical Myc promoter sequence. The "default" clone is the one recommended by SwitchGear. Before deciding which reporter vector(s) to choose, we performed a DNA sequence alignment analysis using NCBI's "blastn" multiple sequence alignment tool (http://www.ncbi.nlm.nih.gov/), to compare the available LightSwitch promoter sequences to the genomic, cDNA and protein coding sequences of Myc gene (Fig. 2). We then selected two LightSwitch promoter vectors for Myc reporter assay development: the default promoter vector (S719565: 0.93 kb) that contains the −0.6 kb position TBE element and the major transcription start site (TATA box), and a longer promoter vector (S722841:1.9 kb) that contains both −1.2 kb and −0.6 kb TBE elements but extends into Myc coding region which will result in extra amino acids at the N-terminus of the reporter protein. The other four promoter vectors are either redundant or lack the necessary TBEs or transcription start site.

In summary, SwitchGear provides a comprehensive reporter assay system that is simple to implement without the need for cloning or DNA preparation. With their experimentally validated pathway assays and stable pathway reporter cell lines (which contain a stably integrated reporter under the control of a validated human regulatory element), a reporter assay could be quickly set up and optimized for HTS applications. However the reporter vectors with desired promoter sequence may not be readily available from their inventory. In such case a custom request can be placed to have a reporter vector prepared by SwitchGear.

Another source to get ready-made reporter vectors is Addgene, a nonprofit plasmid repository that distributes 45,000 plasmids on behalf of 2000 labs from around the world (https://www.addgene.org/). Many reporter vectors reported in publications are available from Addgene for a nominal fee. However the collection is only available to nonprofit organizations.

1.1.3 Custom-Made Reporter Vectors

To design the custom Myc promoter reporter, we first chose the promoter sequence containing the two TBE elements and Myc transcription start site but not the translation start site (1.5 kb in length, Fig. 1). Since we are interested in comparing transient exogenous reporter assay to the X-MAN NanoLuc-PEST Myc endogenous promoter reporter assay, we chose to use the same reporter NanoLuc-PEST (*see* Subheading 1.1.3) to minimize artifacts from different reporter genes. The 1.5 kb Myc promoter sequence was de novo synthesized and cloned into PNL1.2 [NlucP] expression vector (Promega) to drive the expression of NanoLuc-PEST (pMyc-NanoLuc-PEST).

1.1.4 Design of Endogenous Locus Assay

One limitation of endogenous locus reporter assay is the small assay window (signal to background ratio: S/B) due to weak endogenous promoter activity, so a reporter protein with high sensitivity should be considered for building an endogenous locus reporter. NanoLuc is a novel, small and bright luciferase from the deep sea shrimp *Oplophorus gracilirostris* that has been developed by Promega. When compared to traditional luciferases, such as firefly or Renilla reniformis, NanoLuc is approximately 150-fold brighter and can be detected even at low expression levels. This enables detection of endogenous gene transcription and protein expression.

When studying gene transcription it is ideal to use reporter protein that has short half-life to ensure a good dynamic range and prevent intracellular accumulation. This allows the system to rapidly respond to external signals. The cellular half-life of a reporter protein can be reduced to improve reporter responsiveness by adding protein degradation sequences to the reporter gene. The PEST sequence is a short signal peptide that is involved in targeting proteins for degradation. It has been used in transcriptional reporter assays to reduce the half-life of reporter proteins [2, 26]. Promega has developed a destabilized NanoLuc reporter with a fused PEST sequence, NanoLuc-PEST, which has a reduced protein half-life of about 20 min but still maintains a bright signal due to the enzyme's high specific activity.

Using NanoLuc-PEST as reporter, Horizon Discovery has developed a series of X-MAN NanoLuc-PEST promoter reporter cell lines (HIF1A, MYC, GLI1, etc.), which are suitable for studying the regulation of endogenous gene expression in high throughput assay system. Using rAAV gene editing system, a homologous recombination is performed to place the NanoLuc-PEST reporter gene directly downstream of the start codon of the gene of interest, allowing assessment of the endogenous promoter activity. A LoxP flanked selection marker is included in the reporter gene cassette that will be excised by Cre recombinase treatment once a stable cell clone is selected and established (Fig. 3).

The X-MAN NanoLuc-PEST Myc promoter reporter has a NanoLuc-PEST knocked-into one of the endogenous Myc alleles in HCT116 cells. HCT116 cell lines are favored since they are approximately diploid, unlike most immortalized cell lines. This reporter cell line showed rapid and robust decreases in NanoLuc signal when the cells were treated with Actinomycin D, a general transcriptional inhibitor, for 6 h (Horizon data). We acquired this reporter line from Horizon for in-house evaluation and to compare it with the exogenous promoter reporter assays we developed using SwitchGear's and our custom-made reporter vectors.

1.2 Relative Performance of Endogenous Locus and Exogenous Reporter Assays

At the assay development stage, we compared the performance of Myc endogenous locus reporter to exogenous promoter reporters. X-MAN NanoLuc-PEST Myc promoter reporter HCT116 cells (Myc-HCT116) were treated with Actinomycin (positive control) and DMSO (negative control) for 6 h and Nano-Glo assay performed following Promega's protocol (*see* Subheading 3.2). The baseline signal of the assay (no compound treatment) was very low in 1536-well format (raw luminescence unit (RLU) = ~300), resulted in small assay window for inhibitor screen (DMSO/Actinomycin = ~2.5). When compound CHIR-99021, a GSK3b inhibitor that activates β-catenin pathway, was included in cell treatment, the assay window increased to ~8.0, and Z'-factor > 0.5, which is robust enough for 1536-well HTS [27].

The two SwitchGear pMyc-RenSP reporter vectors (S719565 and S722841) and our custom-made pMyc-NanoLuc-PEST reporter vector were tested in transiently transfected HCT116

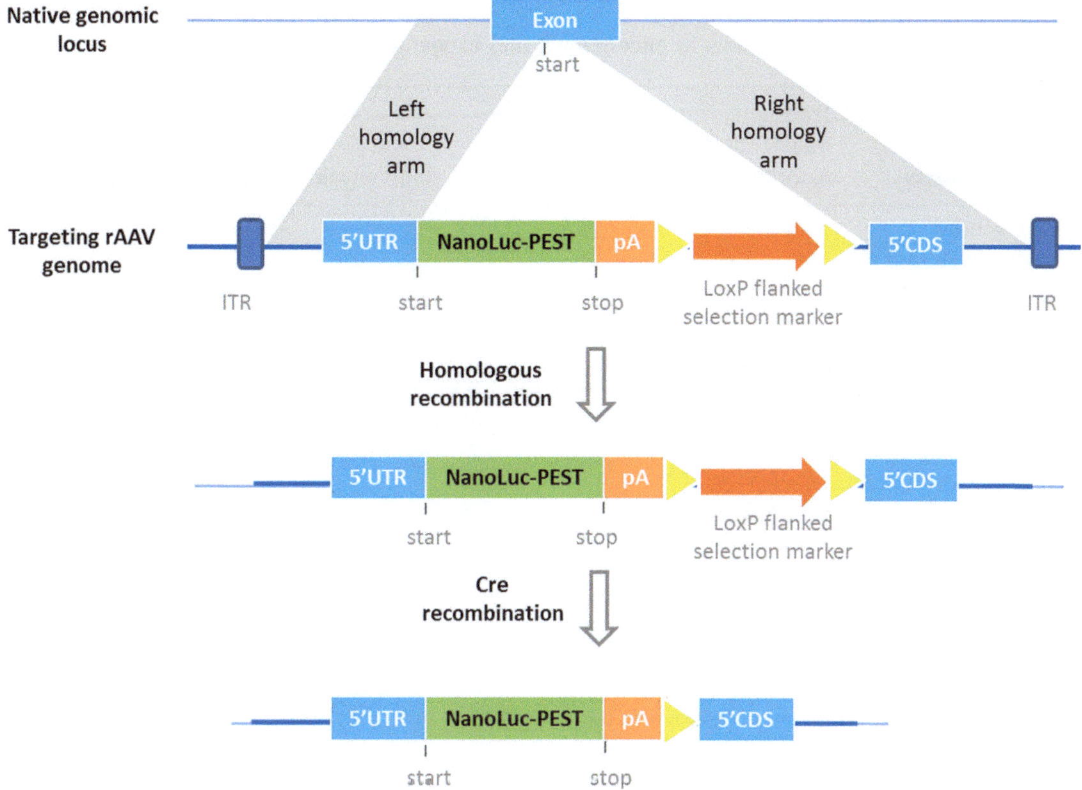

Fig. 3 Generation of X-MAN NanoLuc-PEST promoter reporter cell lines using rAAV gene editing technology (modified from Horizon Discovery's product document). The targeting rAAV genome is designed to include the reporter gene and a mammalian selection cassette flanked by two regions of homology to the target gene. NanoLuc-PEST is placed directly downstream of the endogenous start codon of the gene of interest. A stop codon and polyadenylation signal sequence ensure standalone (i.e., not fused) NanoLuc-PEST is translated. When human somatic cells are infected with rAAV particles, the single stranded viral genomes direct homologous recombination leading to introduction of the reporter gene at the required endogenous locus. Abbreviations: ITR: inverted terminal repeat, UTR: untranslated region, CDS: coding sequences, pA: polyadenylation signal sequence

and HEK293T cells. In both cell lines S722841 didn't generate significant luciferase signal, likely due to the promoter sequence being extended into the endogenous Myc coding region. S719565 (pMyc-RenSP-1) and pMyc-NanoLuc-PEST both showed good assay window (>10) in HCT116 and HEK293 cell lines. However the reporter raw signals were much lower in HCT116 cells even though the transfection efficiency was adequate, resulting in higher variance between wells (coefficient of variation: %CV >30%) in 1536-well plate and thus limiting assays to 384-well format in HCT116 cells. In HEK293 cells the reporter assays were robust (assay window ~15, Z'-factor > 0.5) and amendable to 1536-well HTS.

Table 1
Control compound dose responses in endogenous and exogenous reporter assays

Reporter assay	Transient transfection reporter vector	Co-transfection with activated beta-catenin	Actinomycin	ERKi	MEKi	GSK3bi	BRDi
Endogenous Myc-HCT116 Nanoluc-PEST reporter	None	+	−	−	−	+	−
Exogenous reporter vector in HCT116 cells	pMYC-RenSP-1 (SwitchGear S719565)		−	−			NT
	pMYC-NanoLuc-PEST(custom made)		−	−		+	+
NT	Not tested						
	No change						
+	Increase (> 1.5-fold)						
−	Decrease (< 0.5-fold)						

We also tested these reporter assays against a panel of compounds and reagents known to affect Myc transcription activity. Only the endogenous Myc-HCT116 cells showed responses to a constitutively activated β-catenin mutant, MEK inhibitor (MEKi), GSK3b inhibitor (GSK3bi), and BRD inhibitor (BRDi) as we expected (Table 1). The pMyc-NanoLuc-PEST reporter showed activated response to GSK3bi similar to the endogenous reporter assay, but didn't respond to activated β-catenin mutant, MEKi and BRDi. The SwitchGear reporter pMyc-RenSP-1only responded to Actinomycin and ERKi, and the responses were not specific when compared to the housekeeping gene promoter reporter controls (β-Actin and GAPDH, data not shown).

Based on assay performance and control compound responses in these reporter assays, the decision was made to use endogenous Myc-HCT116 reporter line and pMyc-NanoLuc-PEST transient reporter in HEK 293 cells for 1536-well HTS of a 50,000 small molecule compound library.

Our data indicated that endogenous reporter assay may provide a more accurate system for studying gene transcriptional regulation than the conventional exogenous reporter assays. However, cost has been an issue for making endogenous reporter cell lines using the proprietary rAVV technology and limited its use in regular laboratories. Two other genome editing technologies have gained popularity recently: TALEN (Transcription Activator-Like Effector Nucleases) [11, 12] and CRISPR (Clustered Regularly Interspaced Short Palindromic Repeats) [9, 10, 28]. Both use endonucleases that initiate double-strand breaks (DSBs) at virtually any genomic target sequence, and are used for many applications including gene knockout, transgene knock-in, gene tagging, and correction of genetic defects. Especially CRISPR/Cas9 system, with its high efficiency, ease of use, and low cost, has provided a rapid and affordable route for making site-specific endogenous gene reporter in an investigator's lab [29–31]. Such endogenous reporters could be valuable tools for studying gene transcriptional regulation and screening compounds that modulate a pathway of interest.

2 Materials

1. X-MAN NanoLuc-PEST Myc promoter reporter in HCT116 cells (Myc-HCT116; Horizon Discovery, Cambridge, UK).
2. HEK293T cell line (ATCC, Manassas, VA).
3. RPMI 1640 including 2 mM L-glutamine and 25 mM sodium bicarbonate (Thermo Fisher, Grand Island, NY).
4. Dulbecco's Modified Eagle Medium (DMEM, Thermo Fisher).
5. Fetal bovine serum (FBS; Thermo Fisher).
6. HEPES buffer (Thermo Fisher).
7. Penicillin/streptomycin (100×) (Thermo Fisher).
8. Geneticin (G418; Thermo Fisher).
9. CHIR-99021 (Selleck Chemicals, Houston, TX).
10. OPTI-Modified Eagle Medium (Opti-MEM, Thermo Fisher).
11. FuGENE 6 transfection reagent (Promega, Madison, WI).
12. NanoLuc-Pest expression vector PNL1.2[NlucP] (Promega).
13. Custom-made 1.5 kb Myc promoter and NanoLuc-Pest reporter construct (pMyc-NanoLuc-PEST).
14. LightSwitch reporter gene constructs with Myc promoter S719565 and S722841 (pMYC-RenSP-1, pMYC-RenSP-2; SwitchGear Genomics, Carlsbad, CA).
15. LightSwitch Luciferase assay reagent (SwitchGear).
16. Nano-Glo Luciferase assay reagent (Nano-Glo; Horizon Discovery/Promega).
17. 1536-well black solid bottom tissue culture-treated plates (Corning, Corning, NY).
18. ViewLux uHTS Microplate Imager (PerkinElmer, Waltham, MA).

19. GNF robotic system with fully automated instruments such as liquid dispenser, sample transfer pintool, washer, incubator, and plate readers (GNF Systems, San Diego, CA).

3 Methods

3.1 Endogenous Myc Locus Reporter Compound Screening

1. X-MAN NanoLuc-PEST Myc promoter reporter HCT116 cells (Myc-HCT116) are cultured in RPMI 1640 including 2 mM L-glutamine and 25 mM sodium bicarbonate, supplemented with 10% FBS and 0.3 mg/mL G418 (10% RPMI). Cells are harvested at ~80% confluence, frozen banked (10^7 cells per vial) in large quantities, and stored in liquid nitrogen tank for HTS use (*see* **Note 2**).
2. On the day of screening, thaw frozen Myc-HCT116 cell vials in warm RPMI 1640 media, centrifuge and resuspend in 10% RPMI at 60,000 cells/mL (*see* **Note 3**).
3. Dispense 5 μL/well cells to 1536-well tissue culture-treated black solid plates (3000 cells/well) (*see* **Note 4**).
4. Incubate at 37 °C, 5% CO_2, 93% humidity overnight.
5. Add 1 μL/well of CHIR-99021 at 4.5 μM in 10% RPMI using GNF dispenser (final concentration 0.75 μM). Column 1–4: 10% RPMI, column 5–48: CHIR-99021.
6. Add 30 nL/well testing compounds using GNF pintool (final compound screening concentration: 10 μM) (*see* **Notes 5** and **6**).
7. Incubate for 6 h at 37 °C, 5% CO_2, 93% humidity incubator.
8. Transfer assay plates to room temperature, and incubate for 10 min (*see* **Note 7**).
9. Add 4 μL/well of 2× Nano-Glo reagent following the manufacturer's protocol, incubate for 10 min at room temperature, read luminescence on Viewlux microplate imager using 90 s exposure (2× bin) (*see* **Note 8**).

3.2 Exogenous Reporter Compound Screening

To minimize cell and reagent variation between different assay wells and plates, bulk cDNA transfection is used to transiently transfect reporter vectors in HEK293T cells. The bulk transfection protocol detailed below is very simple to perform offline and increases screen throughput significantly compared to the well-by-well transfection method. The transfected cells are then seeded in 1536-well plates for compound treatment and reporter detection assay. Frozen banked HEK293T cells appear to be sensitive to transfection reagent treatment resulting in poor cell viability and sub-optimum assay performance (lower S/B and z'-factor), so freshly cultured HEK293T cells are used for reporter transfection and HTS.

1. HEK293T cells are cultured in high glucose DMEM supplemented with 10% FBS and 1× penicillin/streptomycin (10% DMEM). Freshly cultured cells are harvested at ~80% confluence on the day of HTS and resuspended in 10%DMEM at 67,000 cells/mL.
2. Reporter vector DNA bulk transfection:
   (a) Mix 450 μL of FuGENE 6 reagent in 36 mL of Opti-MEM (1:80 dilution)
   (b) Add 108 μg of pMyc-NanoLuc-PEST plasmid DNA to FuGENE 6/Opti-MEM dilute, mix well, and incubate for 30 min at room temperature (FuGENE: DNA ratio = 4.2) (*see* **Note 9**).
   (c) Mix DNA/FuGENE 6/Opti-MEM mixture (b) with HEK293T cell suspension [1] at 1:10 ratio, dispense 5 μL/well cells to 1536-well black tissue culture-treated plates (3000 cells/well) (*see* **Notes 3** and **4**).
   (d) Incubate for 24 h at 37 °C, 5% CO_2, 93% humidity.
3. Add 30 nL/well testing compounds using GNF pintool (final compound screening concentration: 12 μM) (*see* **Notes 5** and **6**).
4. Incubate for 6 h at 37C, 5% CO_2, 93% humidity incubator.
5. Transfer assay plates to room temperature, and incubate for 10 min (*see* **Note 7**).
6. Add 4 μL/well of 2× Nano-Glo reagent, incubate for 10 min at room temperature, read luminescent signal on Viewlux microplate imager using 30 s exposure (2× bin) (*see* **Note 8**).

3.3 HTS Data Analysis

The primary screens of a 50,000 compound library using Myc-HCT116 endogenous reporter line and pMyc-NanoLuc-PEST reporter in transiently transfected HEK293T cells were performed in singlicate ($n = 1$). Here we use the endogenous reporter screen as an example to illustrate the HTS data analysis process:

1. Data normalization and QC:

   The raw luminescence unit (RLU) files generated from View-Lux reader were uploaded into a local database where the compound library plate information (compound ID, concentration, etc.) can be associated with screen data. The RLU values of screened samples were then converted to the percentage of inhibition (INH %) based on DMSO control (0% inhibition) and Actinomycin control wells (100% inhibition). The statistics of each assay plate (S/B, CV%, z'-factor) was calculated and plates with S/B < 6 or z'-factor < 0.4 were removed for rerun. The overall health of the screen can be assessed by the statistics of individual assay plates and the overall data distribution (Fig. 4).

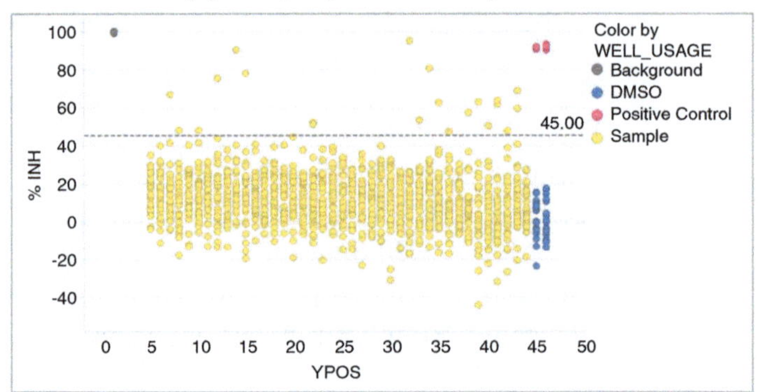

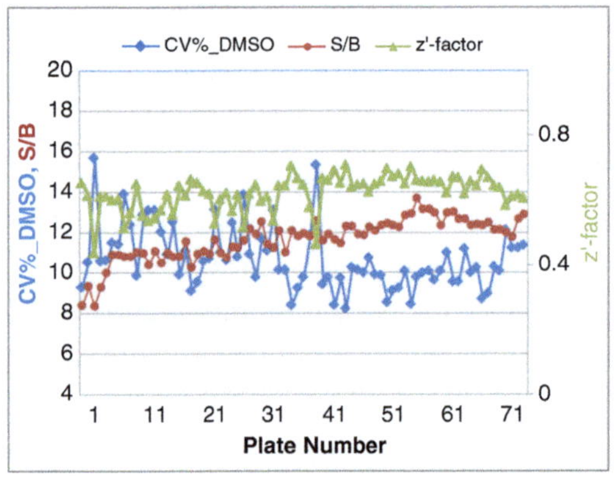

(C) Screen data distribution (%INH)

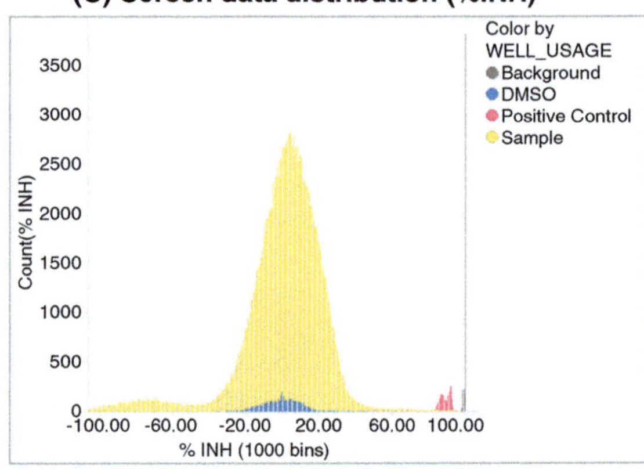

Fig. 4 Primary screening of a 50,000 compound library using endogenous Myc promoter reporter assay. (**a**) Scatter plot of a sample plate from HTS. % Inhibition of samples in a 1536-well plate is shown along the plate column number. Each dot represents a well in 1536-well plate. The samples above the 45% cutoff line represent possible hits that inhibits Myc reporter activity. (**b**) HTS plate statistics of a 75 plate run in the primary screen (partial screen). (**c**) Primary screen data distribution colored by different well usage (control or sample). %INH of all screened sample shows good normal distribution

2. Hit selection:

   To reduce false negative rate and maximize the number of primary hits that can be retested in confirmatory screening, we used a four criteria hit selection strategy that select the top ranking 1% of compounds from each of the four scores: [1] % inhibition (e.g., %INH >45% cutoff), [2] % adjusted inhibition (inhibition adjusted for positional effects), [3] Z-score [32, 33], and [4] B-score [32, 33]. The union of four selection hits were then cherry-picked and rescreened in the same reporter assay in triplicates ($n = 3$). Hits confirmed in triplicate test are followed up in other screen assays (counter screen) and in the same reporter assay in dose titrations (dose response screen).

4 Notes

1. Sequential/nested deletion analysis is a technique used to characterize gene regulation elements such as promoter or enhancer region. For example, a promoter region is first cloned into a vector to drive the expression of a reporter gene that can be measured properly in a reporter assay. To determine the upstream limit of the promoter, a series of nested deletions could be created where the 5′ limit of the promoter is successively closer to the transcription start and their reporter activities are analyzed. Deletions of sequences that are not important for regulation would not affect promoter activity whereas deletion of an important region would lead to deregulation.
2. Frozen cells offer a big advantage in HTS applications compared to freshly cultured cells. It is easier to scale up the cell culture in advance of HTS campaign and be ready for any schedule change of robotic timeline (resource saving). Cells are prepared and banked in bulk at similar culture condition and passage number and can be quality checked in advance to avoid any cell quality issues arising from live culture. The frozen cell preparation should be validated during assay development to ensure comparable assay performance, cell stability, and DMSO tolerance as those of fresh cultured cells.
3. Cell stability should be tested during assay development or validation stage to determine the frequency of cell drops or bulk transfections during HTS. Compound and reagent stability should be tested as well and replenish as needed to ensure assay performance.
4. For a luminescent assay that produces low signal, white solid bottom plates may be used to increase raw signal intensity, as they offer maximum reflection of light. However it may add nonspecific signal from "cross talk" which occurs when light from one well travels through the well walls into adjacent wells. Black plates may be helpful in reducing well-to-well cross-talk but can reduce the raw signal by absorbing some of the light produced by the assay. From our experience we found that 1536-well black solid bottom plates produced the best assay window and z'-factor compared to white solid bottom and black clear bottom plates. The HTS was performed using black solid bottom plates, with a few black clear bottom plates included in each run to observe cell growth condition.
5. Compound collection to be screened are placed in the center of assay plates (e.g., column 5–44), various screening controls such as positive (Actinomycin) and negative (DMSO) control samples are placed in the wells on the side columns (e.g.,

column 1–4 or 45–48). If possible it's best to avoid the edge wells in column 1 and 48 for critical controls that are used for screen data normalization or final data calculation, as the edge wells in cell-based assay tends to have "edge-effect" due to evaporation in higher density plates in extended period of culture.

6. Compound screening concentration is usually optimized during HTS validation run (to validate assay protocol on robotic system). Different compound addition volumes are tested to check cell viability and assay performance. A couple of sample plates randomly picked form a compound collection can be tested to estimate hit rate and help deciding compound screening concentration.
7. Nano-Glo and other luciferase detection assays are sensitive to temperature. Incubation of assay plates at room temperature to equilibrate the plate before adding detection reagent can help by maintaining the same detection condition and thus stabilizing assay window across plates.
8. ViewLux detection settings can be optimized during assay development to achieve better sensitivity and assay window for different assays. For example, the endogenous reporter assay has low RLU and longer ViewLux exposure time (90 s) is chosen to maximize assay sensitivity, whereas the exogenous reporter assay has much higher RLU value, so the ViewLux exposure time is shortened to 30 s to avoid saturation of luminescent signal. Alternative plate readers, including photomultiplier tube based instruments, can also be adjusted for gain and integration time to achieve appropriate signal linearity.
9. Reporter vector transient transfection protocol is optimized during assay development in selected cell lines using control reporter vectors such as pCMV-GFP, pGAPDH-RenSP, and pCMV-NanoLuc. The selections of transfection reagent (lipid), lipid–DNA ratio, lipid–DNA incubation time, cell density, and transfection method (forward vs. reverse transfection) are determined for each cell line based on transfection efficiency, cell viability and reporter assay performance (raw signal intensity, S/B ratio, z'-factor).

References

1. Feng Y, Mitchison TJ, Bender A, Young DW, Tallarico JA (2009) Multi-parameter phenotypic profiling: using cellular effects to characterize small-molecule compounds. Nat Rev Drug Discov 8:567–578
2. Fan F, Wood KV (2007) Bioluminescent assays for high-throughput screening. Assay Drug Dev Technol 5:127–136
3. Chiba T, Tsuchiya T, Mori R, Shimokawa I (2012) Protein reporter bioassay systems for the phenotypic screening of candidate drugs: a mouse platform for anti-aging drug screening. Sensors (Basel) 12:1648–1656
4. Durocher Y et al (2000) A reporter gene assay for high-throughput screening of G-protein-coupled receptors stably or transiently

expressed in HEK293 EBNA cells grown in suspension culture. Anal Biochem 284:316–326

5. Alam J, Cook JL (1990) Reporter genes: application to the study of mammalian gene transcription. Anal Biochem 188:245–254
6. Bronstein I, Fortin J, Stanley PE, Stewart GS, Kricka LJ (1994) Chemiluminescent and bioluminescent reporter gene assays. Anal Biochem 219:169–181
7. Schenborn E, Groskreutz D (1999) Reporter gene vectors and assays. Mol Biotechnol 13:29–44
8. Tuupanen S et al (2009) The common colorectal cancer predisposition SNP rs6983267 at chromosome 8q24 confers potential to enhanced Wnt signaling. Nat Genet 41:885–890
9. Doudna JA, Charpentier E (2014) Genome editing. The new frontier of genome engineering with CRISPR-Cas9. Science 346:1258096
10. Mali P et al (2013) RNA-guided human genome engineering via Cas9. Science 339:823–826
11. Zhang F et al (2011) Efficient construction of sequence-specific TAL effectors for modulating mammalian transcription. Nat Biotechnol 29:149–153
12. Uhde-Stone C, Cheung E, Lu B (2014) TALE activators regulate gene expression in a position- and strand-dependent manner in mammalian cells. Biochem Biophys Res Commun 443:1189–1194
13. Topaloglu O, Hurley PJ, Yildirim O, Civin CI, Bunz F (2005) Improved methods for the generation of human gene knockout and knockin cell lines. Nucleic Acids Res 33:e158
14. Khan IF, Hirata RK, Russell DW (2011) AAV-mediated gene targeting methods for human cells. Nat Protoc 6:482–501
15. Liu X et al (2004) Targeted correction of single-base-pair mutations with adeno-associated virus vectors under nonselective conditions. J Virol 78:4165–4175
16. Inoue N, Hirata RK, Russell DW (1999) High-fidelity correction of mutations at multiple chromosomal positions by adeno-associated virus vectors. J Virol 73:7376–7380
17. Miller DG et al (2005) Large-scale analysis of adeno-associated virus vector integration sites in normal human cells. J Virol 79:11434–11442
18. Munoz IM, Szyniarowski P, Toth R, Rouse J, Lachaud C Improved genome editing in human cell lines using the CRISPR method. PLoS One 9:e109752
19. Dang CV (2012) MYC on the path to cancer. Cell 149:22–35
20. Dang CV (1999) c-Myc target genes involved in cell growth, apoptosis, and metabolism. Mol Cell Biol 19:1–11
21. Gearhart J, Pashos EE, Prasad MK (2007) Pluripotency redux--advances in stem-cell research. N Engl J Med 357:1469–1472
22. Lin CY et al (2012) Transcriptional amplification in tumor cells with elevated c-Myc. Cell 151:56–67
23. He TC et al (1998) Identification of c-MYC as a target of the APC pathway. Science 281:1509–1512
24. Morin PJ et al (1997) Activation of beta-catenin-Tcf signaling in colon cancer by mutations in beta-catenin or APC. Science 275:1787–1790
25. Korinek V et al (1997) Constitutive transcriptional activation by a beta-catenin-Tcf complex in APC−/− colon carcinoma. Science 275:1784–1787
26. Li X et al (1998) Generation of destabilized green fluorescent protein as a transcription reporter. J Biol Chem 273:34970–34975
27. Inglese J et al (2007) High throughput screening assays for the identification of chemical probes. Nat Chem Biol 3:466–479
28. Zhang F, Wen Y, Guo X (2014) CRISPR/Cas9 for genome editing: progress, implications and challenges. Hum Mol Genet 23:R40–R46
29. Rojas-Fernandez A et al (2015) Rapid generation of endogenously driven transcriptional reporters in cells through CRISPR/Cas9. Sci Rep 5(9811)
30. Hisano Y et al (2015) Precise in-frame integration of exogenous DNA mediated by CRISPR/Cas9 system in zebrafish. Sci Rep 5 (8841)
31. Li M et al (2017) Establishment of reporter lines for detecting fragile X mental retardation (FMR1) gene reactivation in human neural cells. Stem Cells 35:158–169
32. Brideau C, Gunter B, Pikounis B, Liaw A (2003) Improved statistical methods for hit selection in high-throughput screening. J Biomol Screen 8:634–647
33. Malo N, Hanley JA, Cerquozzi S, Pelletier J, Nadon R (2006) Statistical practice in high-throughput screening data analysis. Nat Biotechnol 24:167–175

Chapter 13

High-Content Reporter Assays

Erica Cook, Jeffrey Hermes, Jing Li, and Matthew Tudor

Abstract

While luminescent reporter gene assays allow for a rapid and relatively interference free assessment of the activation state of a luminescent reporter, fluorescent reporters do not. They suffer from artifacts such as compound fluorescence and cellular debris which makes the assessment of whole well fluorescence signals difficult. However, the use of high-content screening allows for the isolation of individual cells, segmentation and thus enables the screener to utilize fluorescent reporters to assess the activation state of such a high-content reporter on a cell by cell level, thus minimizing artifacts. Here we discuss the use of such a high-content reporter that enables screening for compounds useful for HIV reactivation on Jurkat cells with high-content screening.

Key words HTS, HCS, High-content screening, Reporter gene assay, HIV, Lentivirus, Latency, Reactivation, High-throughput screening, Drug discovery, 1536-well plates

1 Introduction

Reporter gene technology is commonly used during the drug development process to monitor cellular events associated with signal transduction and gene expression [1]. Traditional reporter gene assays are based on reporter genes which are transiently or stably transfected into host cells harboring the pathway of interest [2]. Expression of the reporter gene is controlled by cis-regulatory elements recognized by sequence specific transcription factors responsive to the pathway's activity. Monitoring transcriptional activation downstream of receptor-mediated signal transduction cascades is an example of reporter gene utility in facilitating the measurement of cellular phenotypes [1]. A key feature of these assays is that activity of the reporter can then be easily measured by means of their enzymatic activity or biophysical feature such as fluorescent light, as an indicator of transcriptional activity. Some of the most commonly used reporter genes include β-galactosidase, luciferase, secreted alkaline phosphatase (SEAP), and green fluorescent protein (GFP) [1].

Robert Damoiseaux and Samuel Hasson (eds.), *Reporter Gene Assays: Methods and Protocols*, Methods in Molecular Biology, vol. 1755, https://doi.org/10.1007/978-1-4939-7724-6_13,

1.1 Types of Reporters

β-galactosidase is a well-characterized bacterial enzyme that has been one of the most widely used reporter genes to monitor transfection efficiency. The main advantage of β-gal reporter gene assays is the assay readouts can be simple and colorimetric or chemiluminescent, but there can be interference from endogenous activity in some mammalian cells [3].

Luciferase is an enzymatic reporter gene that catalyzes the oxidation of various substrates such as luciferin and coelenterazine, resulting in light emission. The most commonly used luciferases for reporter gene assays are bacterial luciferases, the firefly luciferase, and Renilla luciferase. Luciferase reporter assays are well characterized, show no endogenous activity, and have a broad dynamic range. However, luciferase readouts do have a disadvantage of requiring addition of a substrate, and in some cases cell lysis. Exceptions to this are Gaussia [4], NanoLuc [5]), which are measured in the supernatant or can enter the cell, respectively. As with all enzymatic reporters, the susceptibility to enzyme inhibitors [6] and stabilizers common in small molecule screening libraries is quite high. In order to mitigate the effects of such compounds, the screening triage funnel may include additional counterscreening assays downstream of the primary reporter screen. These screening assays may use the same reporter enzyme but under the control of an alternative promoter, or a time course of luciferase activity. Another approach is an orthogonal screen where the same promoter is used in the same cell background but a different reporter enzyme is used to report on activity. The latter approach is effective as it confirms biology of the compound independent of assay technology.

Green fluorescent protein (GFP) has been widely used as quantification does not require cell lysis or additional substrates for the generation of green light. It provides an excellent means for monitoring gene expression and protein localization in living cells and is also a convenient indicator of transfection, allowing cells to be sorted by fluorescence-activated cell sorting (FACS) [7]. Originally discovered in the jellyfish *Aequorea victoria*, GFP is a naturally fluorescent monomeric protein that is stable in the presence of denaturants and proteases, as well as over a broad range of pH and temperatures [8]. Random or site-directed mutagenesis at amino acid residues around the chromophore can change the color and intensity of GFP's fluorescence [9]. Other varieties of *Aequorea* GFP include blue, sapphire, cyan, and yellow [10]. The availability of different colored GFPs allows multiple proteins tagged with distinct spectral variants to be visualized in cell simultaneously [10].

Due to the auto fluorescence of cell debris, dust, and culture media, not to mention some chemical compounds being assayed, GFP assays are ideally read out by imaging rather than whole-well readouts such as via plate readers. Imaging allows segmentation of individual cells, removal of artifacts such as cell clumps or debris, and facilitates multiplexed measurement of viability as well as morphological characterization.

The use of fluorescent protein such as GFP in cellular studies has allowed researchers to track the localization and trafficking of proteins directly in living cells using time lapse fluorescence microscopy [10]. This feature eliminates the need to fix and stain cells with antibodies specific to a protein of interest. Pairs of GFPs can also be used to study protein-protein interactions in living cells using fluorescence resonance energy transfer (FRET), a phenomena whereby light that is emitted from one fluorophore will directly excite the second fluorophore if the two are closer than a minimum distance apart. Blue fluorescent protein (BFP) and green fluorescent protein (GFP) or cyan fluorescent protein (CFP) and yellow fluorescent proteins (YFP) are common FRET pairings that can be used [10]. Additionally, fluorescent protein complementation assays have been devised, whereby two parts of a GFP are independently nonfluorescent, but when brought together by the interaction their fusion partners, become fluorescent [11].

1.2 Benefits of High-Content Reporter Gene Assays

The benefits to using high-content reporter gene assays are numerous. Biologically speaking, reporter gene assays have the advantages of low background activity in cells but amplify the signal from the cell surface to produce a highly sensitive and detectable response [1]. Reporter genes can be transfected into a variety of cell types depending on the transfection methods used, from immortalized mammalian cell lines to suspension cells and primary cell lines [12]. These systems can also be used in combination to provide multiple readouts from a single microplate well (multiplexing) in addition to the measurement of simultaneous cellular events [13]. Events that can be measured simultaneously include cellular proliferation, toxicity events, apoptosis, cell cycle, and also secondary biomarkers that may be of interest to the particular target or pathway being measured. Some reporter genes also afford the ability of a kinetic readout, measuring gene expression or signal transduction over a period of time [1, 13].

In general, the major advantage of cell-based reporter gene assays is their adaptability for high-throughput screening (HTS) strategies. When coupled with the availability of cell lines expressing biologies of interest, these reporters can be used to identify and measure the activation of specific signal transduction pathways in highly sensitive, nonradioactive, streamlined functional assays [1]. The development of reporter proteins and detection methods provides sensitive readouts from intact cells which has facilitated the development of reporter cell lines tailored to HTS requirements. These assays can easily be miniaturized to run on automated robotic platforms to test millions of compounds of potential therapeutic interest to the pharmaceutical industry.

1.3 Reporter Gene Assays: Cell Types and Transfections

When developing a gene reporter assay, the most important consideration will be the experimental application. This will dictate the type of cell line one will want to use in the assay as well as the need to generate a stable or transient cell line. The type of cell line that is needed dictates the type of transfection method that will need to be used to deliver DNA to the cell line of interest. The generation of stably transfected cell lines is essential for a wide range of applications including gene function studies and screening assays. Stable expression allows long term, defined, and reproducible expression of the gene of interest. However, these cell lines can be time-consuming and cumbersome to generate, as stably transfected clones typically have to be selected and cultured over weeks or months. In addition, the stability of expression must be monitored and cell passage limits established to avoid phenotypic drift from the originally derived clone. Transient transfection can be an option for amenable cell lines and can be transfected "on the fly" for HTS screening as another option in assay development. However, transient transfection can lead to greater variability within the cell population, and also additional batch-to-batch variability. This inherent characteristic might be especially problematic when scaling up for HTS. In many cases though, HTS assays can be optimized for transient transfection to achieve acceptable statistical metrics and repeatability sufficient to identify active agents (small molecules, siRNA, etc.) [14].

Optimization of DNA delivery into the cell line selected is necessary and is dependent on the cell line chosen. Different mammalian cell lines have distinct optimal methods for introducing foreign DNA, and there can be a large degree of variability in transfection conditions [15]. The single most important factor in optimizing transfection efficiency is selecting the proper transfection protocol. Common techniques that are currently used include lipid-mediated transfection, electroporation, and viral transduction [12, 15]. The choice of transfection method determines which cell types can be targeted for stable integration of the reporter construct. Lipid-mediated transfection can be used to transfer DNA into many adherent mammalian cell lines. However, delivery of DNA can be difficult in suspension cells lines and primary cell lines, for which alternative methods such as electroporation or viral transduction must be used.

1.4 HTS Case Study

In this chapter, we present a case study in which a miniaturized HTS assay was run using a GFP reporter gene assay. This high-content screen was multiplexed with two additional viability markers to provide a multiplexed readout in one assay plate. Initially, a primary ultrahigh-throughput screen (uHTS) assay was designed using an HIV Latency model system based on a Jurkat T-cell line that utilizes a luciferase reporter under the control of the HIV LTR [16]. The LTR is an HIV retroviral promoter that flanks the

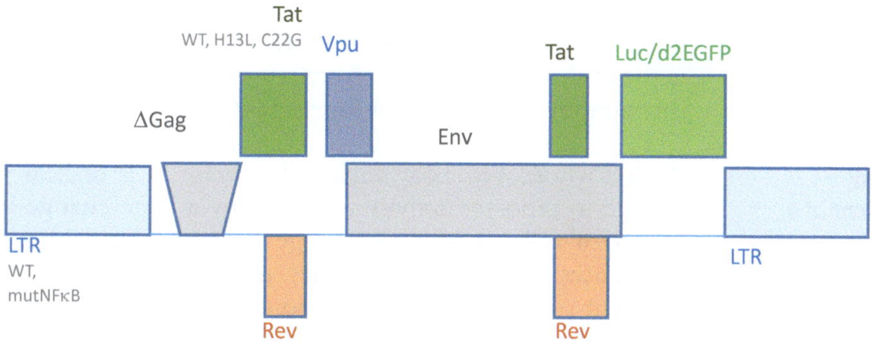

Fig. 1 Genomic organization of the lentiviral vectors used to generate Jurkat-eGFP clones. Components of the lentiviral vector include both genes and gene products as well as genomic structural elements [20, 21]. From left to right: LTR: Long terminal repeat, the DNA sequence flanking the genome of integrated proviruses. Important, as it contains regulatory regions such as those for transcription initiation. Gag: Genomic region encoding the capsid proteins. A deletion was engineered to remove part of the Gag-pol polyprotein in order to render the provirus replication-incompetent. Tat: Transactivator of HIV gene expression. One of two essential viral regulatory factors for HIV gene expression. Rev: The second necessary regulatory factor for HIV expression. Vpu: Viral protein U, unique to HIV-1. Type 1 integral membrane protein. Env: Viral glycoproteins

genome of the integrated proviruses. It contains important regulatory regions, specifically those for transcription initiation. In this model system, stable clones were selected such that the HIV LTR is inactive under normal growth conditions, but can be reactivated by compounds that have been observed to induce latent HIV expression in primary cell models [17–19]. A description of the design of the plasmid construct can be found in Pearson et al., and a schematic is shown in Fig. 1 [16]. Compounds identified as hits in a primary screen of >1 million compounds were retested in dose response format in the presence or absence of submaximal concentrations of reagents known to activate HIV gene expression in the model system. Luciferase activity was the readout of the primary screen and was run in parallel with a viability assay measuring ATP levels. In an effort to characterize the active compounds further, an additional triage campaign was run in five Jurkat T-cell model systems containing a GFP reporter under the control of the HIV LTR [16]. These models differed in integration site locus (by virtue of being independently derived random proviral insertions), as well as engineered Tat and Nef mutations, and various integration multiplicities. The various reporter lines were selected to display a range of sensitivity to latency reversing agents. The assay was miniaturized into 1536-well format and multiplexed to include a Hoechst stain for nuclear count and a propidium iodide stain as an indicator of toxicity. High-content imaging cytometry (Acumen Explorer, TTP Labtech, Cambridge, UK) was used with the assay to screen a total of 300,000 wells. Using the hierarchical clustering features available in Spotfire (TIBCO Software, Palo Alto, CA), compound signatures based on activity were created across the five cell lines so that the compounds of interest could be easily identified. Compounds that exhibited HIV reporter activation in multiple cell lines without significant toxicity were prioritized for follow up in lower-throughput primary T-cell experiments.

1.5 Assay Development of a Reporter Gene Assay

With a whole-well assay such as luciferase, it is unclear if a given change in reporter activity is caused by a large change in a small subset of cells or a small change in the majority of cells. Such differences are important if they translate to the clinic, yielding different pharmacodynamic responses. Since there was interest in differentiating low-level but broadly expressed phenotypes from infrequent large ones, it was of interest to develop an imaging-based (high content) reporter assay. Imaging assays permit numerous characteristics to be cataloged for every cell, thus yielding a "high-content" readout with cellular resolution. The various parameters measured permit nuances in cell response from treatment to treatment, and from cell to cell within a treatment, to be measured and compared. Cell phenotypes can occur on a continuous gradient, or in a dichotomous manner, with different frequencies, and combinatorially. High-content imaging assays permit the characterization of population responses whereas whole-well single parameter readouts do not.

Compared to the primary screen, in which a single readout, luciferase activity, was reported, developing a follow-up assay with a GFP reporter assay afforded the ability to add additional endpoints to measure multiple events from each well. When one is developing this type of high-content reporter gene assay, there are several things to consider at the assay development stage which will help guide the design and execution of the assay. One major decision is the number of samples which need to be screened. This will dictate the assay plate format as well as the imager or cytometer used to acquire images of the samples. Another decision to be made is what other endpoints would be useful to measure as secondary readouts for the screen. Such secondary readouts could serve for normalization, or for identification of undesirable responses such as stress or toxicity.

Depending on lab/facility infrastructure, screens of <10,000 samples are feasible in 384-well plates with a traditional high-content imager, flow cytometer, or a plate-based imaging cytometer. A traditional high-content microscope offers greater resolution and is better suited for detection and analysis of subcellular phenotypes, than an imaging cytometer. If endpoints are only aimed at identifying a particular population or overall intensity of the stain, a plate-based imaging cytometer will work as well. These instruments trade lower resolution for higher acquisition speed, and are consequently best suited for phenotypes where cell intensity changes are of interest. Overall, the assay plate format and imager selected will impact the time it takes to run a study, in addition to instruments needed for liquid handling, and the amount of consumables required. In our case study, we needed to rapidly screen over 1500 compounds in a dose response format across multiple cell lines, so a 1536-well miniaturized assay was deemed necessary. Given high well count, plate read time became an issue that needed to be considered. This consideration, as well as the convenience of a live cell assay, led us to settle on a plate-based cytometer, the Acumen Explorer, to image the wells of the 1536-well assay plate. The Acumen Explorer is a plate-based imaging cytometer that is well suited for HTS of high-content reporters where the readout is based on overall intensity of a particular channel and/or the percentage of cells in a population responding to treatment. The Acumen Explorer can scan a 1536-well plate in about 5 min per laser scan. For this particular assay, two additional endpoints were added to the GFP readout in this assay to provide counterstains. Hoechst 33342 nucleic acid stain is a cell-permeant nuclear counterstain that emits blue fluorescence when bound to dsDNA. It can be added to a live or fixed cell population to stain the nuclei of all cells. Since the dye stains all cells, whether live or dead, it can be used to count the total number of cells in the well and can also be used for normalization purposes when quantifying the percentage of a population that may be responding. Furthermore, the intensity of Hoechst staining correlates to DNA content and can thus be used to measure cell cycle distribution, and responses that affect cell cycle (e.g., by blocking at one of the checkpoints), can be used to identify compounds inducing stress without necrosis. A second stain, propidium iodide, was used to identify frank toxicity. Propidium iodide (PI) is a red-fluorescent DNA stain that is not permeant to live cells. Upon cell death, the cell membrane is compromised, allowing the dye to stain the nucleus.

To set this type of acquisition on the Acumen Explorer, two laser scans are necessary:

1. 405 nm laser excitation and emission data is collected in the 460 nm channel for the Hoechst stain.
2. 488 nm laser excitation and emission data collected in the 530 nm channel for GFP and the 617 nm channel for propidium iodide.

Since the Acumen Explorer performs one laser scan at a time, two separate images are created and analyzed separately, not as a composite. Once assay plate format and counterstains have been chosen, assay development can begin. Assay development for a high-content reporter assay is not much different from other cell based assays. One key experiment is the optimization of cell density. An optimized cell density needs to be chosen such that the image analysis software can effectively segment individual cells in the image, while also providing sufficient cells from which to collect data.

Several assay development experiments are required to optimize the additional endpoints and stains used in the assay. Titrations of both the Hoechst and propidium iodide stains are necessary to ensure the staining intensity of each is high enough to be detected over the background, yet not over-saturated, where it may become toxic to cells or cause difficulty in segmenting individual cells. In the case of live cell assays where cell health may change subsequent to (or in response to) staining, a time course of incubation should be performed to identify the minimum staining time required to obtain a stable signal. Finally, in cases where one excitation wavelength is used to excite and emission is measured from more than one fluorophore at different emission wavelengths, one should consider spectral carryover between the emission wavelengths. In the case of spectral multiplexing, two emission wavelengths overlap and one emission spectrum carries over into the second detection channel. This can be avoided by appropriately titrating the counterstains, as well as selecting appropriate filters and image acquisition parameters that provide low background intensities of the emission wavelengths. In the case of our assay, we had to ensure the GFP signal from the cells was not contributing carry over in the PI channel, as both channels were excited with the 488 nm laser. Optimization of this kind can be accomplished by including wells that are not counterstained with PI, and measuring the background intensity that is generated by the GFP expression alone in the 617 nm emission wavelength, as well as the reciprocal (PI-stained, GFP negative) control.

Additional assay development experiments that need to be included may relate to the specific biology of the assay. Specific examples include optimization of the concentrations of any activators or agonists, and optimization of compound screening concentration. The latter may be chosen based on a pilot screen of one or a few compound plates, with the goal of identifying conditions in which most treatments are nontoxic and the hit rate (frequency of compounds with desired phenotype), is suitable, e.g., 0.1–5%. In the case study presented here a critical experiment was the determination of the concentration of a histone deacetylase inhibitor (HDACi) to use in each of the Jurkat clones that was to be screened. Since one of the goals of the screen was to find compounds that would synergistically activate HIV in the presence of a known class of latency reversing agents, the screen was conducted in the presence of a concentration of HDACi that induces a submaximal increase in HIV transcription. Thus molecules that enhance that effect could be identified. The practical difficulties in maintaining a low but nonzero level of reporter activation required careful monitoring of cell response from day to day.

Finally, if the high-content reporter assay is going to be run on a fully automated robotic platform, additional stability studies should be conducted to ensure a smooth transition to the robotics platform. One should consider factors such as the length of time it will take to sequentially dispense cells into assay plates, and ensuring the cell suspension will be remain viable from beginning to end. Another consideration is to ensure assay reagents are stable in the buffers or media in which they are diluted for the length of time they will be on the robotic system.

1.6 HTS of a Reporter Gene Assay

Following assay development of the reporter gene assay, a screen can commence either on a fully automated robotics platform, or in a semiautomated mode using liquid handlers and plate readers in batch mode. The following portion details the materials and methods used to execute the reporter gene case study.

2 Materials

1. Jurkat-LTR-eGFP clones (generous donation from Jonathan Karn, Case Western University).
2. RPMI (−) phenol red (Thermo Fisher).
3. Fetal bovine serum (heat-inactivated) (Sigma).
4. Pen-Strep (100×).
5. GlutaMAX 1 (Thermo Fisher).
6. HDACi (e.g., Trichostatin A, Selleckchem).
7. Hoechst 33342 (Thermo Fisher).
8. Propidium iodide (Thermo Fisher).
9. 1536-Well black, clear bottom, PDL-treated plates (Aurora Biosciences, San Diego, CA).
10. Acumen Explorer EX3 (TTP Labtech, Cambridge, MA).
11. GNF robotic system with fully automated instruments including liquid dispensers, sample transfer pintool, washer, incubator, plate readers (GNF Systems, San Diego, CA).

3 Methods

1. Cells are passaged and the assay is run in RPMI/5%FBS (HI) with GlutaMAX™ and penicillin/streptomycin. The cells are passaged 2× per week at 1:5 and 1:10 dilutions. A less concentrated passage is prepared the day before the assay so that the cells are around 1×10^6/mL (*see* **Note 1**).

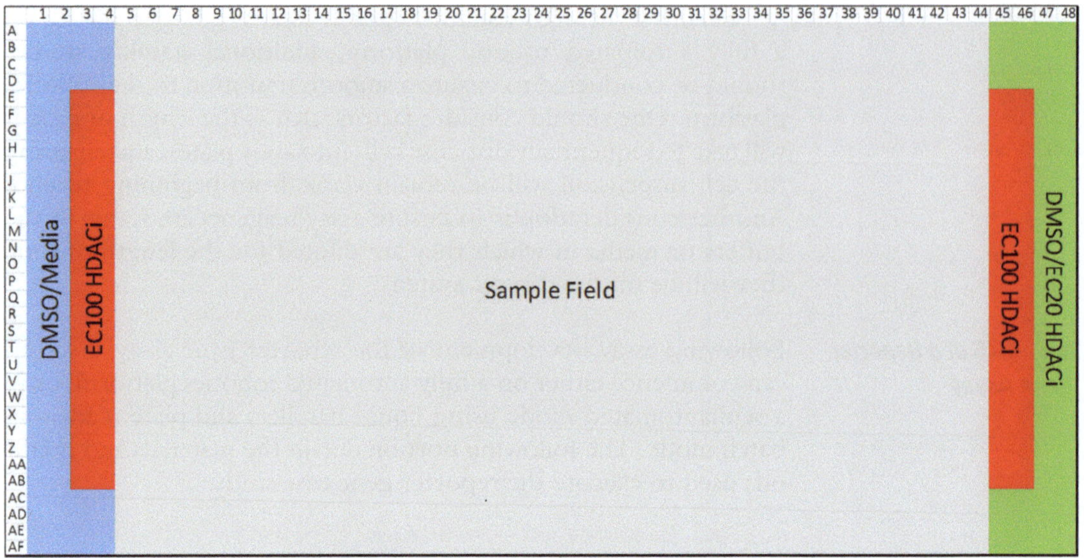

Fig. 2 1536-well platemap for HTS screening of HIV-LTR-eGFP reporter gene assay

2. On the day of the assay, 2100 cells are plated in 2 μL of media in 1536-well, black, clear-bottom PDL treated assay plates (*see* **Note 2**).
3. One microliter of assay media or 1 μL of the EC_{20} of HDACi is added to the plate according to the plate map in Fig. 2.
4. Ten nanoliters of sample compound is added to assay plates using a pintool for sample transfer.
5. Plates are incubated overnight in humidified CO_2 incubators.
6. The next day, Hoechst and propidium iodide dyes are diluted in assay media. 3 μL of dye solution is added to the assay plate well so that the final concentrations of Hoechst and propidium iodide dyes are at 1 and 10 μM final concentrations, respectively.
7. The assay plates are incubated for 50 min at room temperature and scanned on the Acumen Explorer (*see* **Note 3**).

3.1 Image Analysis and Data Normalization

The image analysis portion of this particular GFP reporter assay can be carried out using standard image analysis software that might be provided with the instrument chosen to acquire the images or any third party software. The basic steps are outlined:

1. Identify nuclei by size.
2. Calculate intensity values for eGFP and PI channels.
3. Use object intensity cutoffs to define each cell (defined by nuclear stain), as positive or negative for either GFP or PI fluorescence.

4. Report on the percentage of cells that are GFP positive and also PI positive.
5. Report on the overall GFP intensity per cell.

The Acumen Explorer software is similar to flow cytometry software, where the user can select populations of cells from which population characteristics can be derived (*see* **Notes 4–6**). Once the image analysis algorithm is configured, the instrument can be run in an "on the fly" analysis mode, whereby the images scanned are automatically analyzed and data blocks exported. Figure 3 details the well scans that were collected from the Acumen Explorer, the population characteristics that can be derived from each channel, and the data normalization that is performed on the population characteristics to generate the final parameters used for secondary analysis. From the 405 laser scan, the number of cells can

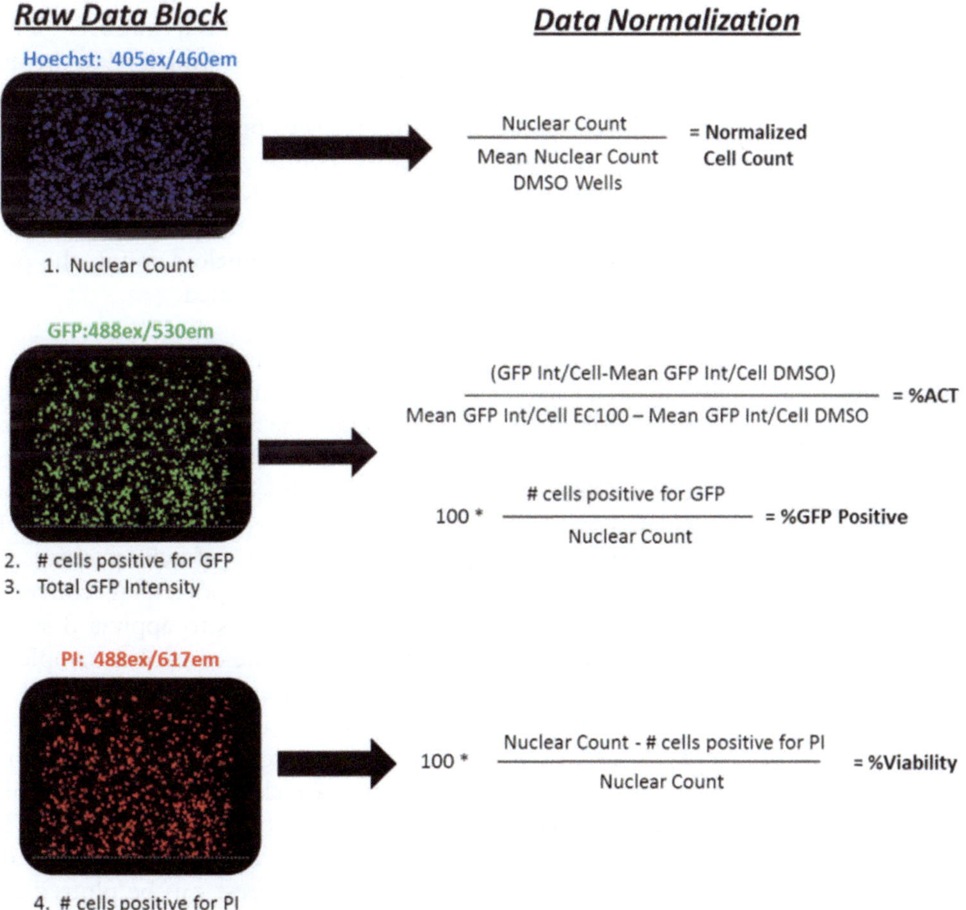

Fig. 3 Four raw data blocks are exported from the analysis routine in the Acumen Explorer software. Additional details show the data normalization that is performed to generate parameters used for secondary analysis including normalized cell count, %Activity, %GFP Positive, and %Viability

be counted based on the nuclear stain. Using the software, the nuclei can be selected by area, using a minimum area of 5 μm and a maximum area of 50 μm. This type of selection eliminates large clumps of cells as well as fluorescent artifacts and provides a total nuclear count. From the 488 scan, a similar analysis was written to identify objects in both the 530 nm channel and 617 nm channel. Once all the GFP objects have been identified, the total intensity of these objects per cell can be derived. Additional analyses include thresholding objects based on a mean intensity to identify cells as positive or negative for GFP expression and propidium iodide staining, and exporting these cell population counts (e.g., total count, GFP positive count, PI positive count).

Additional data normalization of the data blocks exported from the software can be completed in third party software. For this assay, a few extra normalizations are required to generate the parameters that will be used for analysis and hit selection. Total nuclear count needs to be normalized to DMSO treated wells to provide the final well measurement of "normalized cell number." The total intensity of all the GFP objects should be normalized to the nuclear count and then normalized once again to the range of EC_{100} of HDACi wells and the DMSO wells to produce the %Activity measurement. Using GFP positive counts *versus* nuclear count, the population characteristic "%GFP Positive" can be derived. Likewise, using the counts for cells positive for propidium iodide and subtracting this value from the overall nuclear count, the population characteristic "%Viability" can be defined.

3.2 Secondary Analysis of Data

High-content gene reporter assays may be used to screen compounds in a variety of formats. The format of the samples screened and the biology of the assay dictate the appropriate methods to use for hit selection. One example of this is when compounds are tested at a single dose in one biological protocol. In this case, one or two calculated parameters can be used to select active compound "hits," for example based upon the percentage of cells positive or the percent activity of the high-content reporter. A conventional method for selecting active compounds is to apply a 3-standard-deviation threshold from the mean of the control or sample fields. One advantage of running a high-content reporter assay is that one can assess compound toxicity from the normalized cell count alone, or as in the case of our assay, a positive stain for cell death such as propidium iodide. Typically, hits are chosen that meet certain criteria for the reporter and additional criteria for toxicity. TIBCO Spotfire is a useful program to use for this type of multiparameter hit selection, as two parameters of interest can be plotted against each other and samples marked that meet the desired criteria for each parameter.

When running compounds in dose response format, as in our case study, curve fitting software such as GraphPad Prism, TIBCO

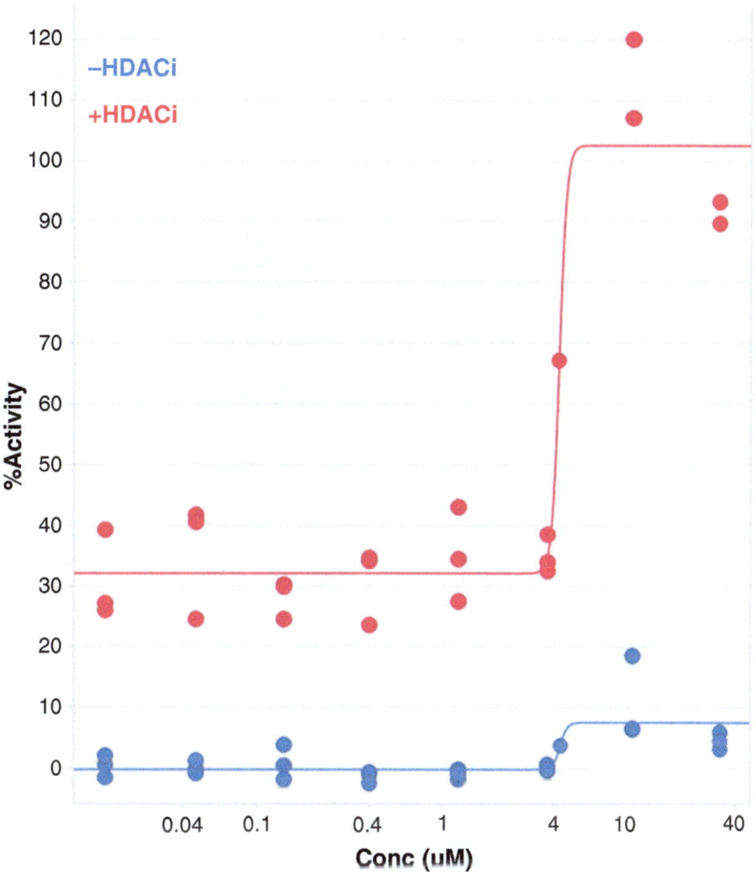

Fig. 4 Sample compound that synergizes with HDACi

Spotfire, or Activity Base should be used to fit dose response curves. It is advisable that automatically generated curves be visually checked and outlying points on the curve should be eliminated from the fitting. These different software packages provide the curve descriptors which include EC_{50} values, E_{max}, E_{min}, and inflection point. In our case study, we can use these curve descriptors to compare changes in EC_{50} and/or E_{max} between the +/– HDACi conditions and make a hit selection cutoff for compounds which show synergy with HDACi by either a change in potency or a change in efficacy. Figure 4 shows a sample compound that has synergistic activity with HDACi. In this case, a significant change in the E_{max} is observed, while there is no shift in potency.

An alternative method for data analysis of synergy screens across multiple cell lines is to use hierarchical clustering of the dose response parameters that are generated for each compound, across each cell line, +/– the synergistic compound. While one could cluster the EC_{50} and/or E_{max} across multiple cell lines and conditions, it was helpful to find a single parameter that could document a change in either of these dose response parameters.

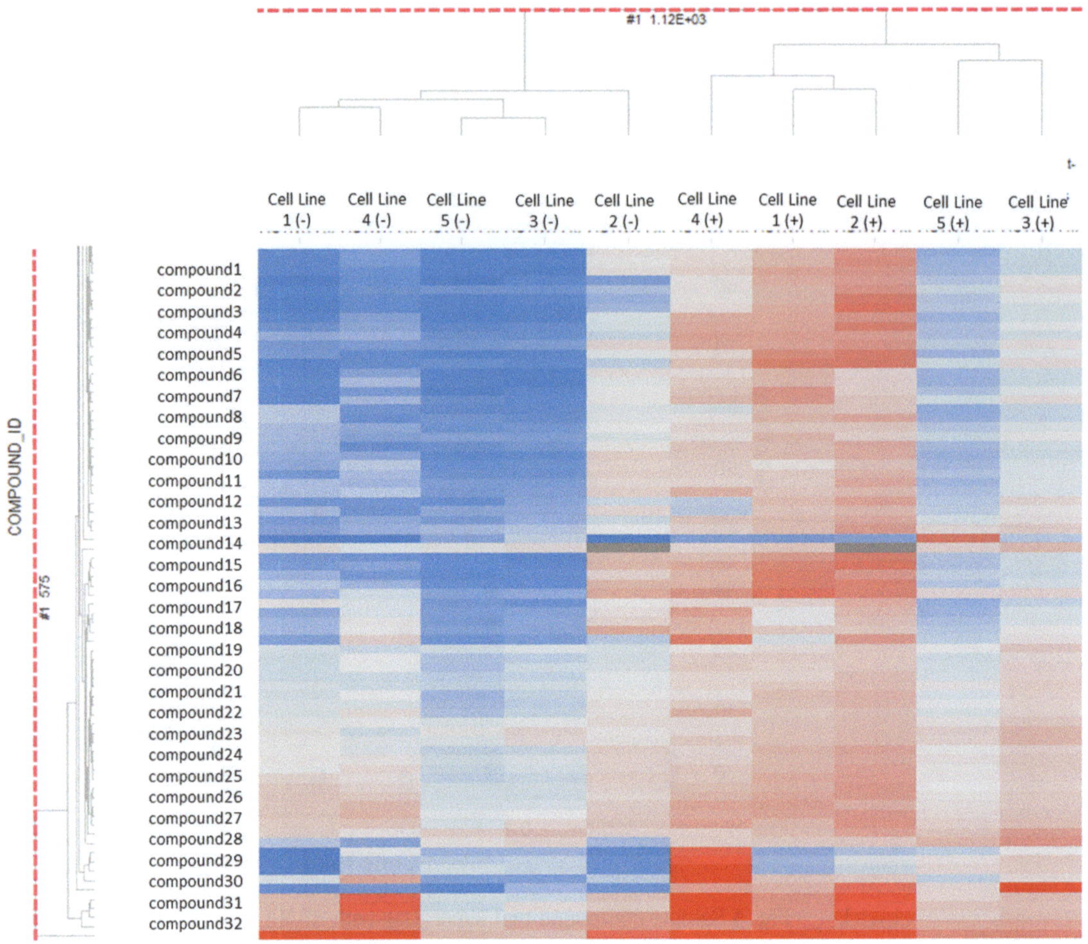

Fig. 5 Hierarchical clustering of the AUC for %Activity generated in the eGFP screen. Individual clones which were not treated with an HDACi ("−") cluster together by activity as do clones which were treated with an EC_{20} concentration of an HDACi ("+"). One can also observe clones that tend to be more reactive or less responsive to the addition of HDACi, or more or less responsive to samples tested alone

We found that area under the curve (AUC) works well for identifying clusters of compounds that show similar response profiles, without having to select strict cutoffs which might lead to missed opportunities (false negatives). Figure 5 shows a sample visualization of this type of clustering which was performed for the eGFP reporter assay described. In addition to the visualization of "compound signatures" across the cell lines, hierarchical clustering allows the cell lines with different genetic backgrounds to be clustered based on compound signatures. This can help to identify the model reporter gene systems that might produce overlapping data and help to eliminate certain cell lines from additional screening. This may also help to identify a reporter gene system that is most predictive of results in a lower throughput, more biologically relevant assay system. Overall, this type of clustering is useful for visualization of the biological activity of a compound across multiple cell lines and conditions, and also for the selection of a list of candidate hits. The candidate list can then be visually inspected to assess the individual dose response curves in order to select compounds with desired activity profiles for follow-up.

3.3 Conclusions

The development of high content-based reporter assays provides researchers with the opportunity to observe cellular dynamics under physiological conditions. By utilizing imaging endpoints, no cell lysis or exogenous substrates are required to detect activity of a high-content reporter gene, which means researchers can study protein localization and transcription over time. These assays can run on live or fixed cells, and as we have shown in the example, they can be multiplexed with other markers to measure effects such as toxicity or other biomarkers of interest. High-content reporters can be transfected into a variety of cell types including immortalized cell lines as well as primary cells lines and suspension cell lines. As we have shown through the example, the main advantage of high-content reporter assays is the ability to miniaturize, multiplex with other readouts, assess cell population heterogeneity, and streamline the protocols so that they can be used in HTS for drug development.

4 Notes

1. Cell stability should be tested during assay development to determine the frequency of cell drops or bulk transfections during HTS. Compound and reagent stability should be tested as well and replenish as needed to ensure assay performance.
2. For high-content assays, black, clear-bottom plates should always be used. The assay plate should be compatible with the high-content reader chosen. For assay plate cytometers, plate flatness is not a major issue. However, for confocal imagers, one will want to select a high quality plate with low bottom thickness variation. If planning to use a higher objective magnification or a water objective, keep in mind that this may limit portions of the plate that can be imaged such as wells along the edge or corners.
3. When running a high-content HTS campaign, image acquisition speed can be an issue. In cases where a high speed cytometer is used or a faster traditional imager, live cell staining followed by rapid image acquisition is feasible, and the whole protocol can be combined in one automated screening run. However, if image acquisition speed is an issue, cells may be

fixed and stained. In this case, the assay plate processing portion of the protocol may be decoupled from the image acquisition portion. Always check that all stains used are compatible with fixation and stable for the length of time it will take to acquire images from all assay plates.

4. In order to optimize scanning speed on the Acumen Explorer, we experimented with resolution and scan area. The Acumen Explorer scans and collect data while moving across the plate. Therefore, only adjustments made in the *Y*-direction will affect speed, such as the resolution in the Y direction and the height of the scan area.
5. The Acumen Explorer imaging cytometer can be engineered with 1–3 lasers. One can select from a 405, 488, 561, and 635 nm lasers for excitation along with various filters to collect data from various fluorophores.
6. Scan settings for each laser/filter should be optimized using a few wells in the assay plate. The wells can be scanned and PMT setting adjusted so that the mean peak intensity is around 1000 (the maximum measurable intensity is 16,383). It should not be much higher or lower. When adjusting scan settings for channels which are related to assay activity, wells that have been treated with a positive control should be used to adjust the PMT settings. A reader protocol can be developed during assay development, but it is important to verify each day the assay is run that the preset parameters still fit the distribution of the controls.

References

1. Naylor LH (1999) Reporter gene technology: the future looks bright. Biochem Pharmacol **58**:749–757
2. Fabio Gasparri AG (2010) Image-based high-content reporter assays: limitations and advanatages. Drug Discov Today Technol **7** (1):21–30
3. Inada A, Nienaber C, Bonner-Weir S (2006) Endogenous beta-galactosidase expression in murine pancreatic islets. Diabetologia **49** (5):1120–1122
4. Tannous BA et al (2005) Codon-optimized Gaussia luciferase cDNA for mammalian gene expression in culture and in vivo. Mol Ther **11** (3):435–443
5. Hall MP et al (2012) Engineered luciferase reporter from a deep sea shrimp utilizing a novel imidazopyrazinone substrate. ACS Chem Biol 7(11):1848–1857
6. Thorne N et al (2012) Firefly luciferase in chemical biology: a compendium of inhibitors, mechanistic evaluation of chemotypes, and suggested use as a reporter. Chem Biol **19** (8):1060–1072
7. Chalfie M, Tu Y, Euskirchen G, Ward WW, Prasher DC (1994) Green fluorescent protein as a marker for gene expression. Science **263**:802–805
8. Patterson GH, Knobel SM, Sharif WD, Kain SR, Piston DW (1997) Use of the green fluorescent protein and its mutants in quantitative fluorescence microscopy. Biophys J **73**:2782–2790
9. Megley CM, Dickson LA, Maddalo SL, Chandler GJ, Zimmer M (2009) Photophysics and dihedral freedom of the chomophore in yellow, blue and green fluorescent protein. J Phys Chem **113**:302–308
10. Tavaré JM, Fletcher LM, Welsh GI (2001) Using green fluorescent protein to study intracellular signaling. J Endocrinol **170**:297–306

11. Yu H, West M, Keon BH, Bilter GK, Owens S, Lamerdin J, Westwick JK (2003) Measuring drug action in the cellular context using protein-fragment complementation assays. Assay Drug Dev Technol **1**(6):811–822
12. Patrick Condreay J, Witherspoon SM, Clay WC, Kost TA (1999) Transient and stable gene expression in mammalian cells transduced with a recombinant baculorvirus vector. Proc Natl Acad Sci U S A **96**:127–132
13. Wolff M, Wiedenmann J, Nienhaus GU, Valler M, Heilker R (2006) Novel fluorescent proteins for high-content screening. Drug Discov Today **11**:1054–1060
14. Unterreiner V, Ibig-Rehm Y, Simonen M, Gubler H, Gabriel D Comparison of variability and sensitivity between nuclear translocation and luciferase reporter gene assays. J Biomol Screen 2009, **14**:59–65
15. Kingston RE, Chen CA, Okayama H, Rose JK (2003) Transfection of DNA into eukaryotic cells. In: Current protocols in molecular biology. John Wiley & Sons, New York
16. Pearson R, Kim YK, Hokello J, Lassen K, Friedman J, Tyagi M, Karn J (2008) Epigenetic silencing of human immunodeficiencty virus (HIV) transcription by formation of restrictive chromatin structures at the viral long terminal repeat drives the progressive entry of HIV into latency. J Virol **82**(24):12297–12303
17. Barton KM, Archin NM, Keedy KS, Espeseth AS, Zhang YL, Gale J, Wagner FF, Holson EB, Margolis DM (2004) Selective HDAC inhibition for the disruption of latent HIV-1 infection. PLOS **9**(8):1–11
18. Rasmussen TA, Tolstrup M, Winckelmann A, Østergaard L, Søgaard OS (2013) Eliminating the latent HIV reservoir by reactivation strategies. Hum Vaccin Immunother **9**:790–799
19. Shirakawa K, Chavez L, Hakre S, Calvanese V, Verdin E (2013) Reactivation of latent HIV by histone deacetylase inhibitors. Trends Microbiol **21**(6):277–285
20. Sundquist WI, Krausslich H-G (2012) HIV-1 assembly, budding, and maturation. Cold Spring Harb Perspect Med 2:1–24
21. Frankel AD, Young JAT (1998) HIV-1: fifteen proteins and an RNA. Annu Rev Biochem 67:1–25

Chapter 14

Data Mining and Computational Modeling of High-Throughput Screening Datasets

Sean Ekins, Alex M. Clark, Krishna Dole, Kellan Gregory, Andrew M. Mcnutt, Anna Coulon Spektor, Charlie Weatherall, Nadia K. Litterman, and Barry A. Bunin

Abstract

We are now seeing the benefit of investments made over the last decade in high-throughput screening (HTS) that is resulting in large structure activity datasets entering public and open databases such as ChEMBL and PubChem. The growth of academic HTS screening centers and the increasing move to academia for early stage drug discovery suggests a great need for the informatics tools and methods to mine such data and learn from it. Collaborative Drug Discovery, Inc. (CDD) has developed a number of tools for storing, mining, securely and selectively sharing, as well as learning from such HTS data. We present a new web based data mining and visualization module directly within the CDD Vault platform for high-throughput drug discovery data that makes use of a novel technology stack following modern reactive design principles. We also describe CDD Models within the CDD Vault platform that enables researchers to share models, share predictions from models, and create models from distributed, heterogeneous data. Our system is built on top of the Collaborative Drug Discovery Vault Activity and Registration data repository ecosystem which allows users to manipulate and visualize thousands of molecules in real time. This can be performed in any browser on any platform. In this chapter we present examples of its use with public datasets in CDD Vault. Such approaches can complement other cheminformatics tools, whether open source or commercial, in providing approaches for data mining and modeling of HTS data.

Key words ADME, Bayesian models, CDD models, CDD vault, Visualization, Collaborative database, Data mining

1 Introduction

For well over 20 years, the early stages of modern drug discovery have utilized high-throughput screening (HTS) of large libraries using small molecules against target-based assays [1]. The approach has also been widely followed by academic screening centers [2–6] which has led to the identification of hundreds of new chemical probes [7, 8]. This approach, while having some failings in several disease areas or targets, most notably antibiotics [9], has led to a

Robert Damoiseaux and Samuel Hasson (eds.), *Reporter Gene Assays: Methods and Protocols*, Methods in Molecular Biology, vol. 1755, https://doi.org/10.1007/978-1-4939-7724-6_14,

shift back to whole cell screening for some disease areas. This also seems to agree with the history of drug discovery in that more drugs were derived from phenotypic approaches [10]. The hit rates of these HTS efforts are usually low and below 1% [11–14]. The resulting data are iteratively analyzed alongside physicochemical properties, cytotoxicity, and any other available data prior to further iterations of testing until an ideal collection of drug candidates is found. This data can also be ultimately used to produce computational models and enable further learning.

Computational approaches have played an increasingly important role in the drug discovery process within large pharmaceutical firms, and now constitute an essential part of drug discovery. Virtual screening of compounds using ligand-based and structure-based methods to predict potency enables more efficient utilization of HTS resources, by enriching the set of compounds physically screened with those more likely to yield hits [15–18]. Computation of absorption, distribution, metabolism, excretion, and toxicity (ADME/Tox) properties exploiting statistical techniques greatly reduces the number of expensive assays that must be performed, and now makes it practical to consider these factors very early in the discovery process to minimize late stage failures of potent lead compounds that are not drug-like [19–25]. Large pharma have successfully integrated these in silico methods into operational practice, validated them, and realized their benefits because these firms have (1) expensive commercial software to build models, (2) large diverse proprietary datasets based on consistent experimental protocols to train and test the models, and (3) extensive computational and medicinal chemistry expertise on staff to run the models and interpret the results. In contrast, drug discovery efforts centered in universities, foundations, government laboratories, and small companies ("extra-pharma") frequently lack these three critical resources and as a result have yet to exploit the full benefits of these in silico methods. As preclinical academic partnerships are important for both the industry as well as universities (in 2015 there were 236 such deals [26]) it will be critical to provide industrial strength computational tools to ensure that early stage pipeline molecules are appropriately filtered before investing in them.

Typical practice in pharma is to integrate in silico predictions into a combined workflow together with in vitro assays to find "hits" that can then be reconfirmed and optimized. The incremental cost of a virtual screen is essentially zero, and the savings compared with a physical screen are magnified if the compound would also need to be synthesized rather than purchased from a vendor. If the blind hit rate against some library is 1% and the in silico model can prefilter the library prospectively, enriching the set of compounds to be tested so the experimental hit rate reaches, say, 2%, then significant resources are freed up to search a broader chemical space, focus more precisely on promising regions, or both [27].

The very high cost of in vivo and in vitro screening of ADME/ Tox properties of molecules is a big motivator to develop in silico methods to filter and select a subset of compounds for testing. By relying on very large internally consistent datasets, large pharma has succeeded in developing highly predictive but proprietary ADME models [19–22]. At Pfizer, as well as other large pharmaceutical companies, many of these models (e.g., volume of distribution, aqueous kinetic solubility, acid dissociation constant, distribution coefficient) [19–22, 28] have achieved such high accuracy that they could be considered competitors to the experimental assays. In most other cases, large pharmaceutical companies perform experimental assays for a small fraction of compounds of interest to augment or validate their computational models. Extra-pharma efforts have not been so successful, largely because they have by necessity drawn upon smaller datasets, in a few cases trying to combine them [25, 29–34]. However, public datasets in ChEMBL [35–38], PubChem [39, 40], EPA Tox21 [41], ToxCast [42, 43], public datasets in the Collaborative Drug Discovery, Inc. (CDD) Vault [44, 45], and elsewhere are becoming available and used for modeling. [46–48].

2 Materials

There have been several efforts describing different data mining [49] and machine learning approaches used with HTS datasets (reporter gene assays, whole cell phenotypic screens, etc.) over the past decade alone, illustrated with the following examples.

2.1 Data Mining Tools

In 2006 Yan et al., published their experiences of data mining from millions of compounds over hundreds of assays. Their development of ontology-based pattern identification was described, as well as scaffold families with structure–HTS relationships, with a focus on finding artifactual results [50]. Crisman et al. took the identification of false positives in reporter gene assays further by building Bayesian machine learning models with 650,000 molecules tested in these assays at Novartis. This resulted in frequent hitter models. These authors also predicted the target families for the frequent hitters, as well as suggested that compounds producing reduced luciferase signals as a readout, such as those compounds inhibiting the cell cycle, would in turn be a source of false positives. Experimental validation was also performed showing an enrichment over random screening [51]. A quantitative HTS screen of a chemical collection against an IL-6 reporter gene assay is typical of many studies, in this case using Leadscope fingerprints and hierarchical clustering, identifying five scaffolds as potential artifacts as they had activity in cells lacking the β-lactamase reporter [52]. Several different machine learning algorithms have been used for data mining including decision tree models; a recent review discussed their use in HTS as well as for ADME/Tox properties suggesting their interpretability was an advantage [53]. Open source Java software called screening assistant 2 [54] was developed for storage and analysis of very large HTS libraries. The use of this software was illustrated on very large libraries of 15 million vendor molecules as well as smaller kinase libraries, for which analysis of physicochemical properties, and scaffold analysis were performed [54]. An example of data mining in large databases is for Basic Active Structures (BAS) that are substructures which are indicative of biological activity [55]. As the number of hits in databases is small, there is a huge imbalance in favor of inactive compounds, which makes it hard to extract substructures of actives. A workflow was demonstrated using random sampling and applied to several datasets such as HIV integrase, HIV-protease, and procaspase-3 from PubChem and MDDR databases [55]. This approach seems very reminiscent of using extended connectivity fingerprints with Bayesian algorithms. Another recent study dealing with the active/inactive imbalance in HTS datasets developed a method called DRAMOTE for data preprocessing which is an active learning approach focused on using HTS data to predict activity [56]. This approach was demonstrated on various PubChem datasets showing improved precision for more than half the datasets [56]. A new database called the BioAssay Research Database (BARD) was described in 2015 for housing screening data for probe development which uses a controlled vocabulary to describe the assay protocols. While this database is free and open-source, it relies on several non-open source components that require licenses [57]. A linked hierarchy visualization in the database shows the distribution of biological process terms for a molecule and allows a linked approach to data mining. It is suggested that the restricted vocabulary used in BARD would make new datasets more powerful [57].

2.2 Visualization in CDD Vault

Processing HTS results is tedious and complex, as the vast amount of data involved tends to be multidimensional, and may well contain missing data or have other irregularities. CDD have recently developed more sophisticated visualization capabilities in CDD Vault which address these problems by providing a suite of modern web-based data visualization tools that generate publication quality data graphics. By making use of a variety of web technologies, including WebGL and SVG, the system allows users to manipulate and visualize hundreds of thousands of points in real time across an arbitrary number of dimensions. A representation of our implementation of this methodology is shown in Fig. 1 in which selected molecules and data are taken from CDD Vault and utilized in the new Visualization module.

We provide a collection of plots (initially scatterplots and histograms) in the main area which shows a general picture of the data.

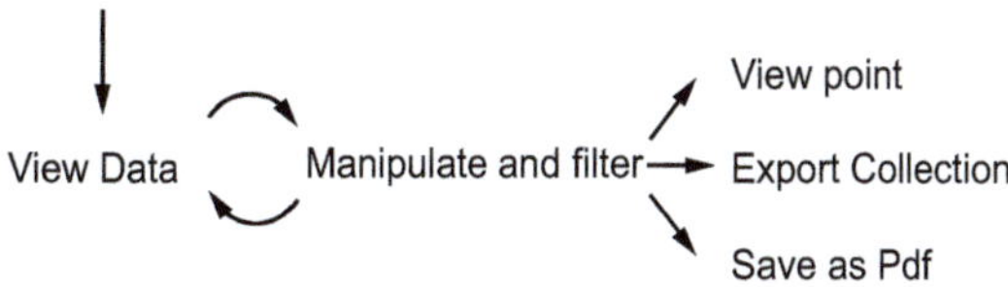

Fig. 1 A flowchart of the user experience flow of the Visualization application in CDD Vault. DOI: https:/doi.org/10.6084/m9.figshare.3206266

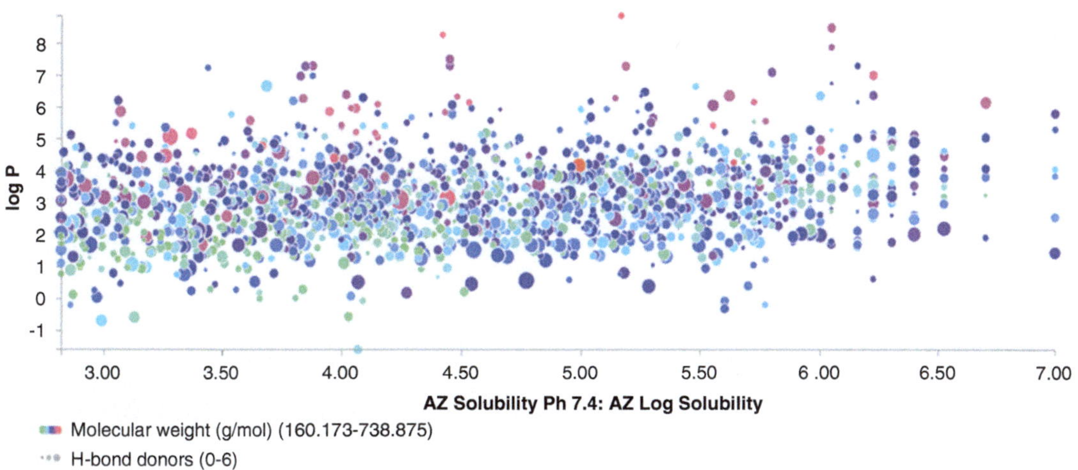

Fig. 2 A sample plot from the Visualization Module in CDD Vault using Astra Zeneca public solubility data from ChEMBL on 1763 compounds showing the relationship with calculated molecular properties. DOI: https:/doi.org/10.6084/m9.figshare.3206266

Much of the power of these plots, in addition to their utilization of the traditional spatial, color, and size variables, comes from their relative awareness of their position in the ambient higher dimensional space.

CDD visualization can be used as follows:

1. For instance, by representing both the selected and unselected points the user is able to visually measure the role of a given point in the larger space, such as a scatterplot utilizing these properties (Fig. 2).
2. The data in these plots can then be manipulated by either adjusting the filters on the right side of the page, or by click and dragging directly on the plots themselves (Fig. 3).
3. If the user wishes to know more about a given molecule or compound they can select it or find it in the data table.
4. Once the user is satisfied with their curation of the plots, they can export their selection to a PDF, or send the collection of molecules back to CDD Vault [44], where it can be used to develop machine learning models using CDD Models [46–48],

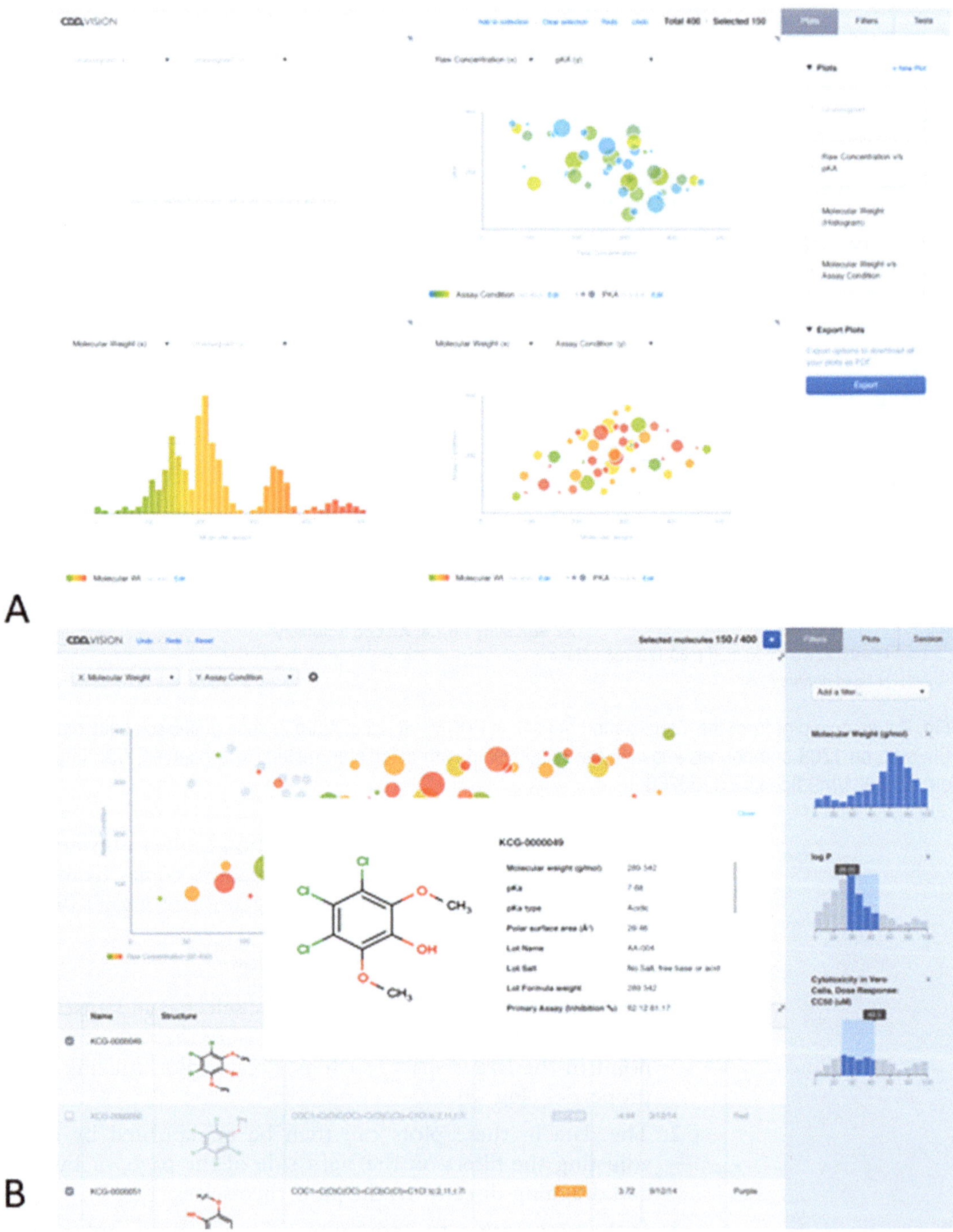

Fig. 3 (**a**) Screenshot of the new Visualization capabilities in CDD Vault, showing The Broad Chagas disease dose response dataset that was used in a recent study by us to build a Bayesian machine learning model [2]. (**b**) A screenshot showing highlighting of structures and filtering of data (right of screen). DOI: https:/doi.org/10.6084/m9.figshare.3206266

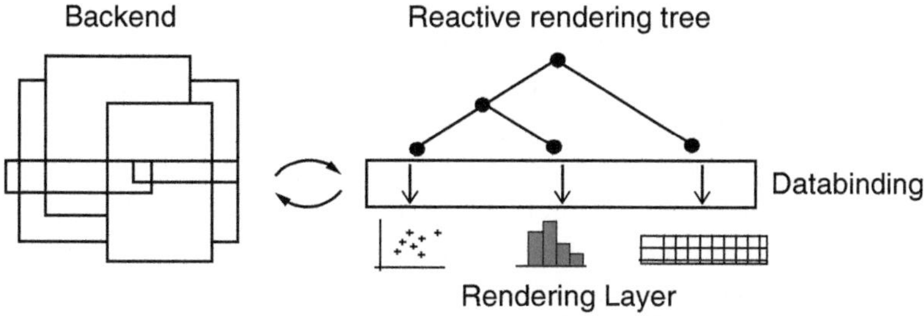

Fig. 4 A flowchart of the technical structure of the Visualization module in CDD Vault. The backend is formed using Immutable and Crossfilter.js, the data binding layer is constructed using d3.js and jQuery, and finally the rendering layer makes use of d3.js and Pixi.js. DOI: https:/doi.org/10.6084/m9.figshare.3206266

or shared securely among other collaborators. In constructing our technology stack we attempted to solve three major problems:

Drug discovery data is often multidimensional and irregular [58], Data exploration is greatly aided by the ability to quickly undo and redo actions, User directed data mining and exploration is most effective when the feedback response loop is minimized to as close to real time as possible.

5. We found that the most effective way to address these constraints in a web context involved implementing a JavaScript application with a reactive architecture (Fig. 4).
6. The CDD Vault Visualization module therefore represents a new tool that allows users to explore and present data via an intuitive user interface. It is a modular platform, and is able to support a wide variety of visualizations with high performance results.

2.3 CDD Models

Recently a novel web-based software capability (CDD Models) was developed within the CDD Vault that enables scientists to work together effectively to discover and improve new drug leads, with the option not to reveal chemical structures to each other. Our goal was to create the first practical system of biocomputational analysis across distributed datasets with different owners, while respecting data privacy, thus lowering the key barrier to collaboration. Using models to accelerate the preclinical drug discovery pipeline will enable groups to *effectively* exploit state-of-the-art computational tools such as bioactivity, ADME/Tox predictions and virtual screening. This will also make it easier for researchers both outside and inside pharma and biotech to collaborate and benefit from high-quality datasets derived from big pharma.

This work was initiated when we collaborated with computational chemists at Pfizer in a proof of concept study which demonstrated that models constructed with open descriptors and keys (CDK + SMARTS) using open software (C5.0), performed essentially identically to expensive proprietary descriptors and models (MOE2D + SMARTS + Rulequest's Cubist) across all metrics of performance, when evaluated on multiple Pfizer-proprietary ADME datasets: human liver microsomal stability (HLM), RRCK passive permeability, P-gp efflux, and aqueous solubility [59]. Pfizer's HLM dataset, for example, contained more than 230,000 compounds and covered a diverse range of chemistry as well as many therapeutic areas. The HLM dataset was split into a training set (80%) and a test set (20%) using the venetian blind splitting method; in addition, a newly screened set of 2310 compounds was evaluated as a blind dataset. All the key metrics of model performance, e.g., R^2, RMSE, kappa, sensitivity, specificity, positive predictive value (PPV) were nearly identical for the open-source approach vs. proprietary software (e.g., PPV of 0.80 vs. 0.82). The open-source approach even computed slightly faster (0.2 vs 0.3 s/compound). All of the datasets studied yielded the same conclusion that models built with open descriptors and models were as predictive as the commercial tools. Following this we proved the value of commercially available Bayesian algorithm and extended connectivity fingerprints in Discovery Studio (Biovia, San Diego, CA) by successfully predicting in advance which subsets of compound libraries our collaborators should screen. Each of the tuberculosis (TB) research groups we collaborated with (Infectious Disease Research Institute, UMDNJ-NJMS and Southern Research Institute) used computational models we created to identify new lead compounds for tuberculosis while saving significant time and expense. The models resulted in prospective potency predictions and achieved screening hit rates of 15–71% for suggested compounds, far higher than the 0.6–1.5% typical for random library HTS screening. This work was subsequently published [60–62] and we extended this approach to lead optimization [63] and dataset fusion alongside evaluation of other machine learning algorithms [64] to evaluate whether bigger models were better [6]. We also demonstrated how machine learning methods could be used to predict in vivo activity in the mouse model for testing for efficacy against Mycobacterium tuberculosis [65]. More recently we have applied the Bayesian machine learning approach to identify leads and repurpose drugs for Chagas disease [66] and Ebola [67].

CDD created a drop-in replacement for the extended connectivity fingerprints of maximum diameter 6 (ECFP6) and molecular function class fingerprints of maximum diameter 6 (FCFP6) fingerprints and the resulting code was made available to the public as a feature in the Chemical Development Kit (CDK) project under an open source license (https://github.com/cdd/modified-bayes). We coded the Bayesian algorithm with these fingerprints and implemented it in CDD Models. This allowed us to deliver a foundational platform for selective, secure distributed model generation and execution, including a toolkit based on open-source algorithms and descriptors. We have applied CDD Models to modeling decision-making for chemical probes [8], ADME-Tox models [9] as well as models of microsomal stability in mouse [68]. The open source descriptors and Bayesian algorithm have also been used outside of CDD Vault to create several thousand Bayesian models with the ChEMBL data [10] as well as Bayesian models for human drug transporters [69] which could be useful for drug discovery and making these models more mobile. This also shows how developing the open source technologies could benefit others outside of CDD and stimulate new technology development. More recently we have developed a Bayesian binning approach which is a step toward semiquantitative Bayesian models [47].

3 Methods

3.1 Applications of Data Mining and Machine Learning in CDD

We have made use of several public datasets to indicate how they can be used in CDD (or elsewhere) for data mining or machine learning with CDD Models and other methods.

3.1.1 Machine Learning Models

1. Bayesian models can be generated using the open source FCFP6 descriptors [70] alone in CDD Models (Collaborative Drug Discovery Inc. Burlingame, CA) [46] and at the same time perform a threefold cross validation.
2. Where possible, we validated the models with an external test set and generated receiver operator (ROC) plots.
3. We have previously described the generation and validation of the Laplacian-corrected Bayesian classifier models developed for various datasets using Discovery Studio 3.5 or 4.1 [71]. This approach has been utilized with the datasets for predicting selectivity. A set of simple molecular descriptors were used: FCFP_6 (Discovery Studio version of this descriptor) [72], AlogP, molecular weight (MW), number of RB, number of rings, number of aromatic rings, number of HBA, number of HBD, and molecular fractional polar surface area were calculated from input SD files.
4. Models were validated using leave-one-out cross-validation in which each sample was left out one at a time, a model was built using the remaining samples, and that model was utilized to predict the left-out sample. Each of the models was internally validated, receiver operator (ROC) plots were generated, and the cross-validated (XV) ROC area under the curve (AUC) calculated. The Bayesian model was additionally evaluated by performing fivefold cross validation in which 20% of the dataset is left out five times.

3.1.2 Abbott Kinase Inhibitors Dataset

1. We have evaluated whether FCFP_6 descriptors and a Naïve Bayesian algorithm could be used to predict selectivity for a large-scale kinome-wide dataset made available by Abbott Laboratories, with data on more than 1487 molecules against 172 kinases [73] available in CDD Vault (https://app.collaborativedrug.com/register).
2. For each compound, a promiscuity value was calculated, defined as the fraction of kinases tested for which the compound had a potency value of 1 μM or less. In the original study [73], the authors found that compounds with more HBD or HBA were more likely to be promiscuous.
3. For our Bayesian model, the 1487 molecules with disclosed structures were used and a cutoff for promiscuity of 0.3 was applied.
4. Using threefold cross validation in CDD Models, this Bayesian model led to a receiver operator characteristic (ROC) value of 0.85 (Fig. 5a).
5. When the model was built with additional simple descriptors as well as FCFP_6 fingerprints (in Discovery Studio), a fivefold cross-validation ROC was equivalent (Fig. 6, Table 1).
6. We then used the kinase selectivity values determined by Ambit for 72 [74] compounds as an external test set to assess the predictions of these models.
7. It was found that when the cut off for selectivity fraction was set at 0.3, the test ROC value was 0.9 (Fig. 5b) when compounds were tested at 300 nM, while the ROC was 0.68 when compounds were tested at 3 μM (Fig. 5b).
8. A model was also built using Discovery Studio which gave a test set ROC of 0.81 at 300 nM (Fig. 6, Table 1).
9. Good kinase selectivity model fingerprint features included nitrogen-rich heterocyclic, substituted anilines, and generally rigid fragments (Fig. 7a). In contrast, the bad kinase selectivity fingerprint features included moieties with multiple positive charges that may reduce selectivity (Fig. 7b). These results suggest that such selectivity models are predictive for external compounds, and could be a useful filter for selecting kinase inhibitors.

3.1.3 Broad 100 Protein Binding Dataset

1. Following the results with the Abbott Kinase dataset, the approaches for predicting selectivity were evaluated for applicability across other proteins.
2. A Bayesian machine learning model was built with more than 15,000 compounds with binding data for 100 different proteins (not just limited to kinases) [75]. This dataset is also available as a public dataset in the CDD Vault.

A

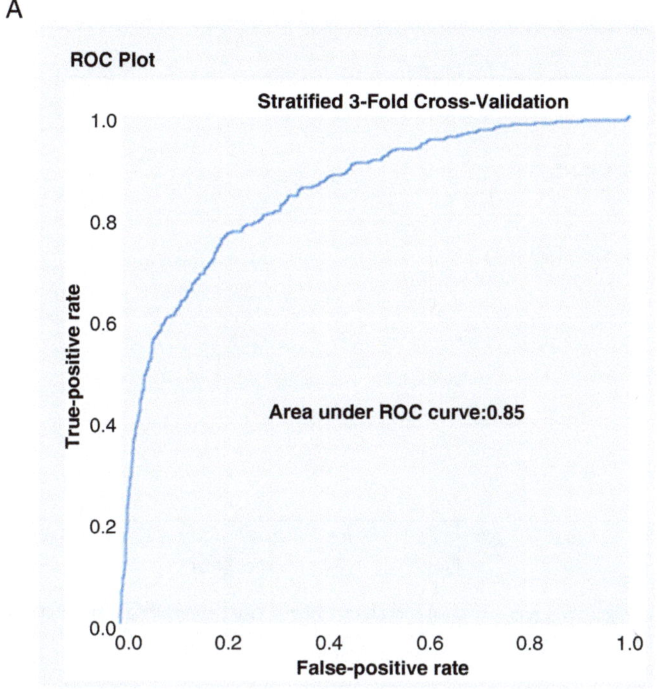

B

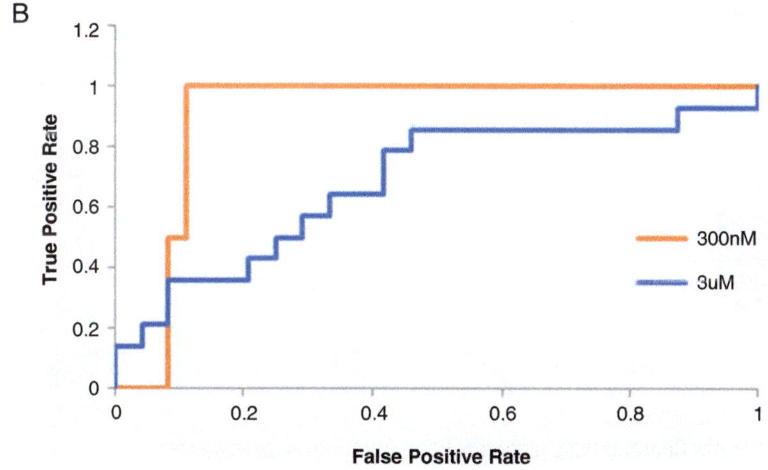

Fig. 5 Receiver Operator Characteristic plots for CDD Bayesian model with FCFP6 descriptors only after threefold cross-validation for predicting selectivity in kinases using Abbott Laboratories data [73]. (**a**) Training set. (**b**) The test set ROC for two different cutoffs using 39 compounds from the Ambit dataset not found in the training set from the Abbot dataset [74]. DOI: https:/doi.org/10.6084/m9.figshare.3206266

3. The cutoff for this model was 0.05 and the resulting threefold cross validation ROC value was 0.78 (Fig. 8). Similar results were obtained using a Bayesian model in Discovery Studio (Fig. 9, Table 2).

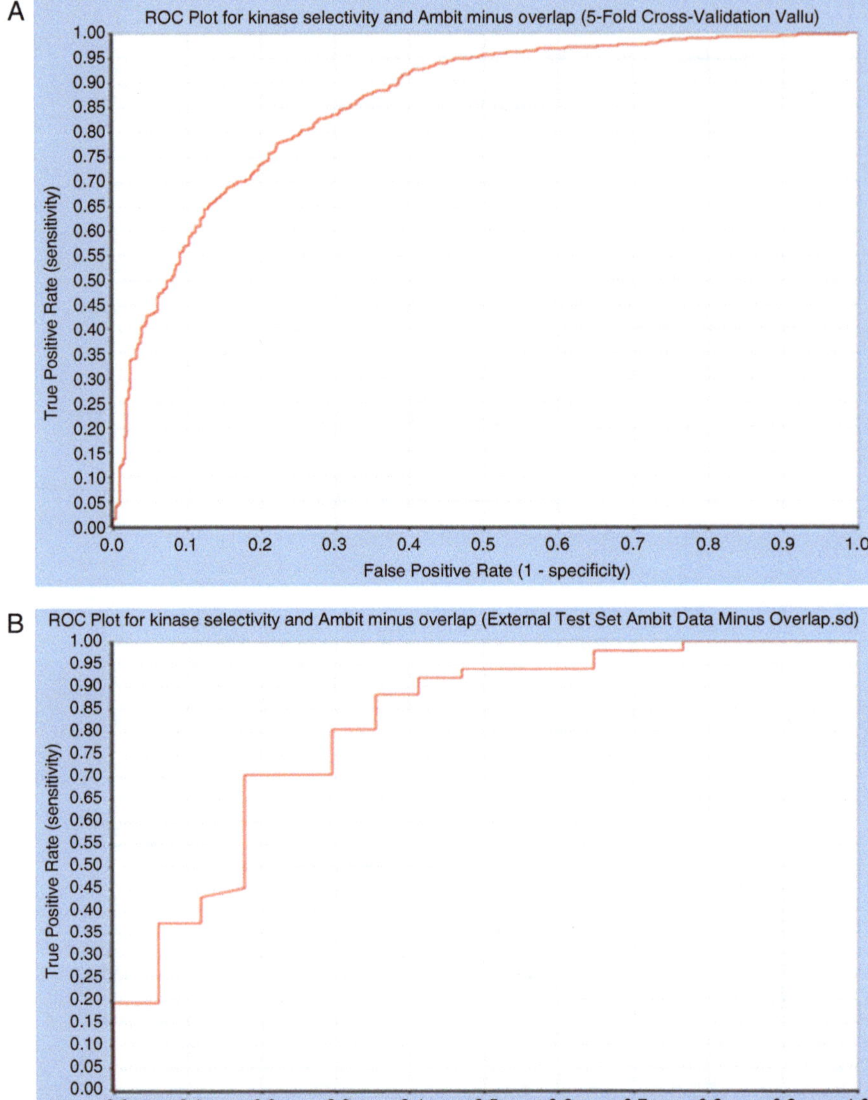

Fig. 6 Receiver Operator Characteristic plots for Discovery Studio Bayesian Models for Kinase Selectivity using Abbott Laboratories data [73]—minus overlapping compounds in Ambit dataset [74]. Descriptors used: ALogP, FCFP_6, Molecular Weight, Number of Aromatic Rings, Number of H-Bond Acceptors, Number of H-Bond Donors, Number of Rings, Number of Rotatable Bonds, and Molecular Fractional Polar Surface Area. Selectivity values less than 0.3 = active. The Ambit dataset was used as a test set after removal of overlapping compounds. (**a**) Training Set. ROC score 0.870 (leave-one-out). Best cutoff for this model is −2.624. (**b**) Test Set ROC = 0.81 (Confusion Matrix: True Positives = 44, False Negatives = 7, False Positives = 6, True Negatives = 11). DOI: https:/doi.org/10.6084/m9.figshare.3206266

4. We have also identified the good and bad fingerprints for this model (Fig. 10) Good fingerprint features include complex natural-product, macrocyclic, and biomimetic functionality.

Table 1
Model statistics for Discovery Studio Bayesian Models for Kinase Selectivity: (A) training set, (B) test set minus overlapping compounds in Ambit

Model name	ROC score	ROC rating	True positive	False negative	False positive	True negative	Sensitivity	Specificity	Concordance
(A) Training set									
Fivefold cross-validation result									
Kinase selectivity and ambit minus overlap	0.858	Good	941	167	37	352	0.849	0.905	0.864
(B) Test set									
Validation result using external test set AmbitDataMinusOverlap.sd									
Kinase selectivity and ambit minus overlap	0.813	Good	44	7	6	11	0.863	0.647	0.809

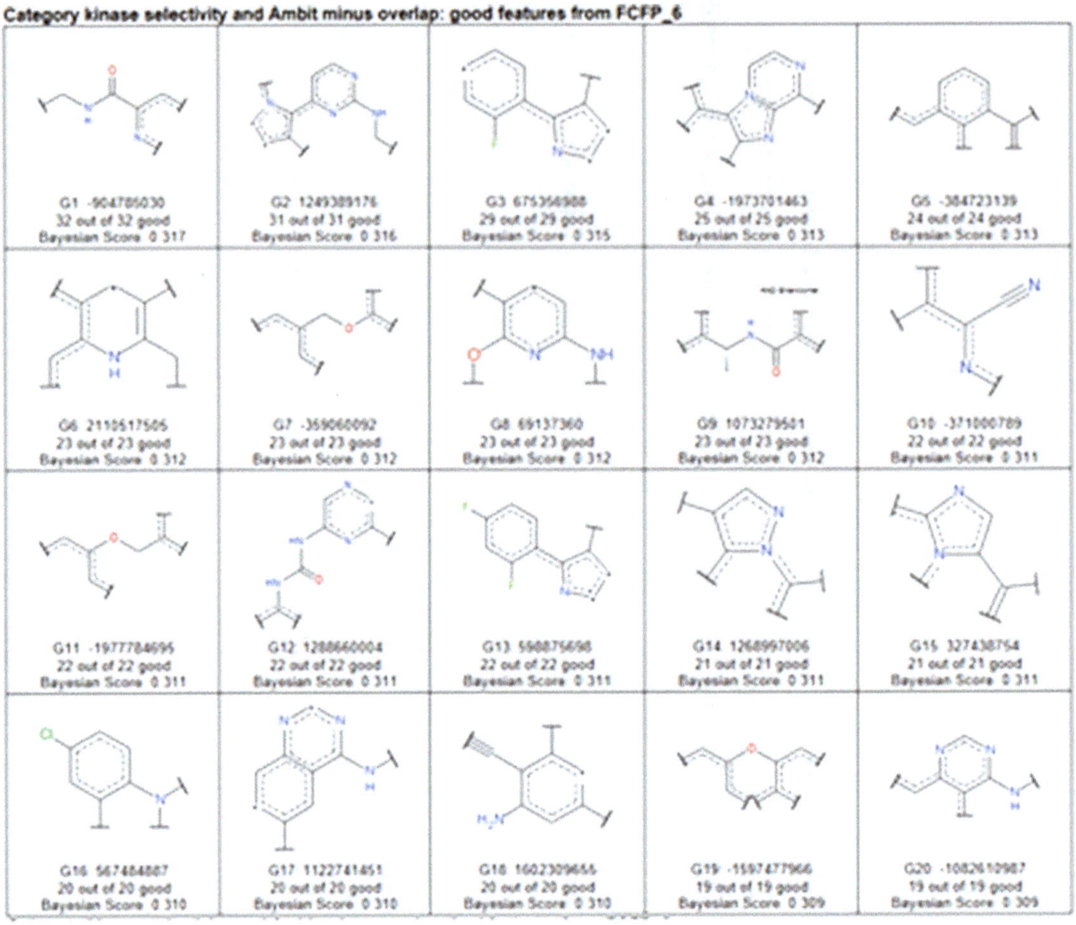

Fig. 7 (**a**) Good Kinase selectivity model good fingerprints. (**b**) Kinase selectivity model bad fingerprints. DOI: https:/doi.org/10.6084/m9.figshare.3206266

In contrast, the bad features include hydrolytically labile functionality, as well as flexible cyclic and acyclic aliphatic functional groups.

5. Together, this suggests that it is possible to glean biological promiscuity trends based on machine learning models using FCFP_6 fingerprints alone or in combination with other simple descriptors. These types of models could also complement other calculations for molecules which we have described recently [76]. More datasets with known drug discovery methods would provide additional confidence in the statistical significance of the results, and further validate the initial trends. For example there has been considerable numbers of studies attempting to understand kinase selectivity [74, 77–80] in order to avoid off target effects and to understand function.

B

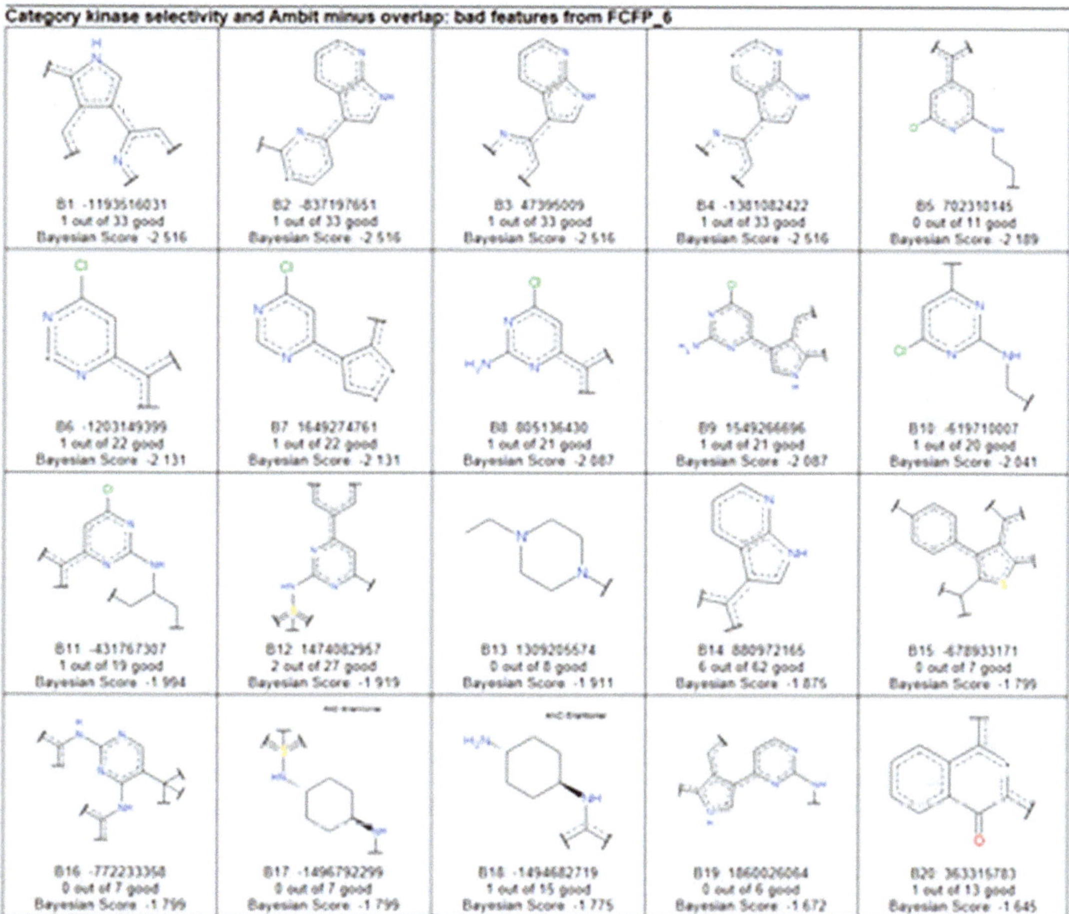

Fig. 7 (continued)

These examples show that learning from some of these datasets can also help predict kinase selectivity. In general, such machine learning approaches can be applied across different proteins and HTS datasets and aid in molecule selection or scoring.

4 Conclusion

In the preceding decade a major transformation has occurred in the pharmaceutical industry which has had to adapt by acquiring companies or partnering with other companies or academic groups to bring in early preclinical innovation or products [81]. We have previously discussed the shift in HTS from industry to academia and the potential bottlenecks and issues this creates [2] around data quality [8, 71]. We and others have also described the need for

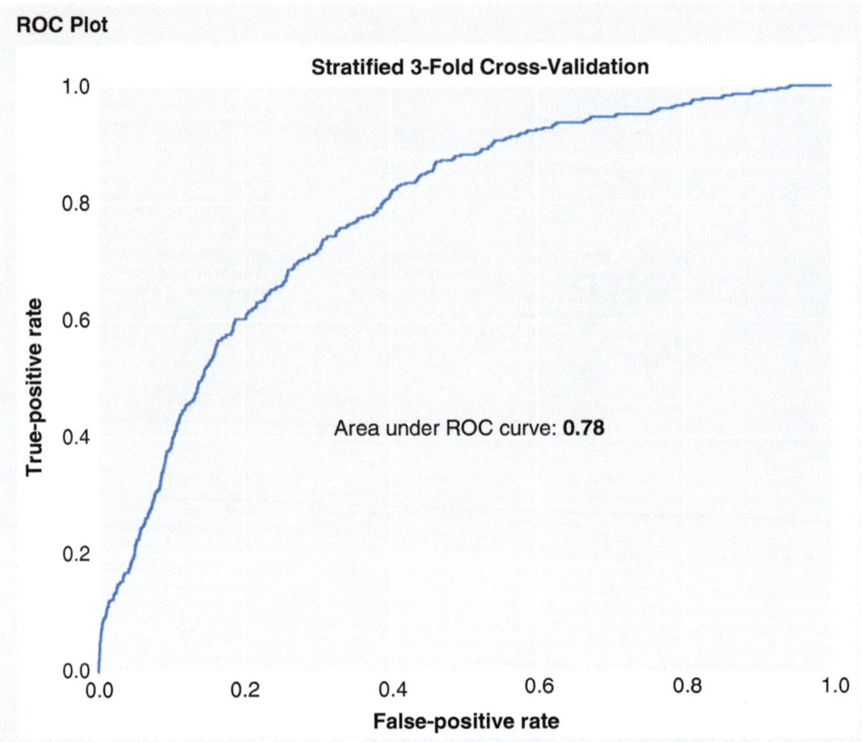

Fig. 8 Receiver Operator Characteristic plot for CDD Bayesian model with FCFP6 descriptors only after threefold cross validation. Promiscuity of compounds binding to proteins using ~15,000 compounds [75] with binding data to 100 different proteins. DOI: https:/doi.org/10.6084/m9.figshare.3206266

informatics for public–private collaborations in the precompetitive space [44, 45, 82] as exemplified by projects such as the NIH Blueprint, Bill and Melinda Gates Foundation TB drug accelerator, Kinetoplastid Drug Development Consortium, and More Medicines for Tuberculosis (Fig. 11). There is also an increasing shift to the cloud for such collaborative projects, e.g., the European Lead Factory [82]. Key issues for all of these efforts (in which there is a big HTS component) is securing intellectual property, masking molecule structures in some cases, and selective sharing with only select members of a consortium.

In the process of developing CDD Vault, we have introduced improved visualization software that enables interactive graphing with multiple plot types, data mining, and publication quality outputs as described for the Visualization Module in CDD Vault. Our approach builds on past ideas on data visualization and mining to produce a scientifically rigorous dashboard for drug discovery data, and suggests that these techniques might also be applicable to any field which has multidimensional data. Such approaches could also be useful for "Big Data" set analysis. The Visualization module in the CDD Vault represents a new tool that is complete unto itself, in

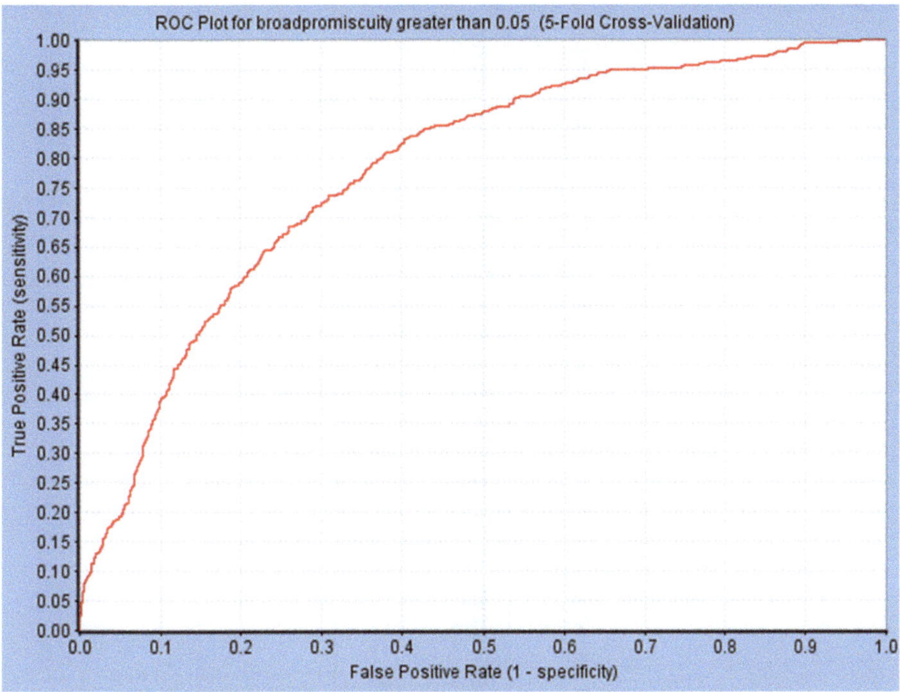

Fig. 9 Receiver Operator Characteristic plot for Discovery Studio Model of promiscuity of compounds binding to proteins using ~15,000 compounds [75] with binding data to 100 different proteins. The following descriptors were used: ALogP, FCFP_6, Molecular Weight, Number of Aromatic Rings, Number of H-Bond Acceptors, Number of H-Bond Donors, Number of Rings, Number of Rotatable Bonds, and Molecular Fractional Polar Surface Area. The cutoff for this model was 0.05. ROC score is 0.784 (leave-one-out). Best cutoff for this model is −0.560. DOI: https:/doi.org/10.6084/m9.figshare.3206266

that it allows users to explore and present data. It is a modular platform, and can support a wide variety of visualizations with ease. Future work includes tools for visually presenting both statistical and chemical clustering in formats including dendrograms and heat maps.

Big pharma has looked at the key chronic diseases in the western hemisphere. Yet if we think about healthcare from a global perspective there are still diseases (e.g., neglected) that are common in the developing world that can in many cases be readily treated with available drugs. Also there are thousands of diseases that occur in small patient populations and are not addressed by any treatments [83]; these are classed as rare or orphan diseases. Neglected and rare diseases traditionally have not been the focus of big pharma, while biotech and academia have been primarily involved in their drug discovery. Our work aims to circumvent these limitations so that extra-pharma discovery projects can benefit from current and emerging best industry software informatics practices. CDD now enables extra-pharma drug discovery projects to benefit from proprietary, commercial ADME/Tox, and other models in

Table 2
Model statistics for Discovery Studio Bayesian Model after fivefold cross-validation for >15,000 compounds with binding data to 100 different proteins

Model name	ROC score	ROC rating	True positive	False negative	False positive	True negative	Sensitivity	Specificity	Concordance
Fivefold cross-validation result									
Broadpromiscuity greater than 0.05	0.778	Fair	366	52	3384	11,444	0.876	0.772	0.775

A

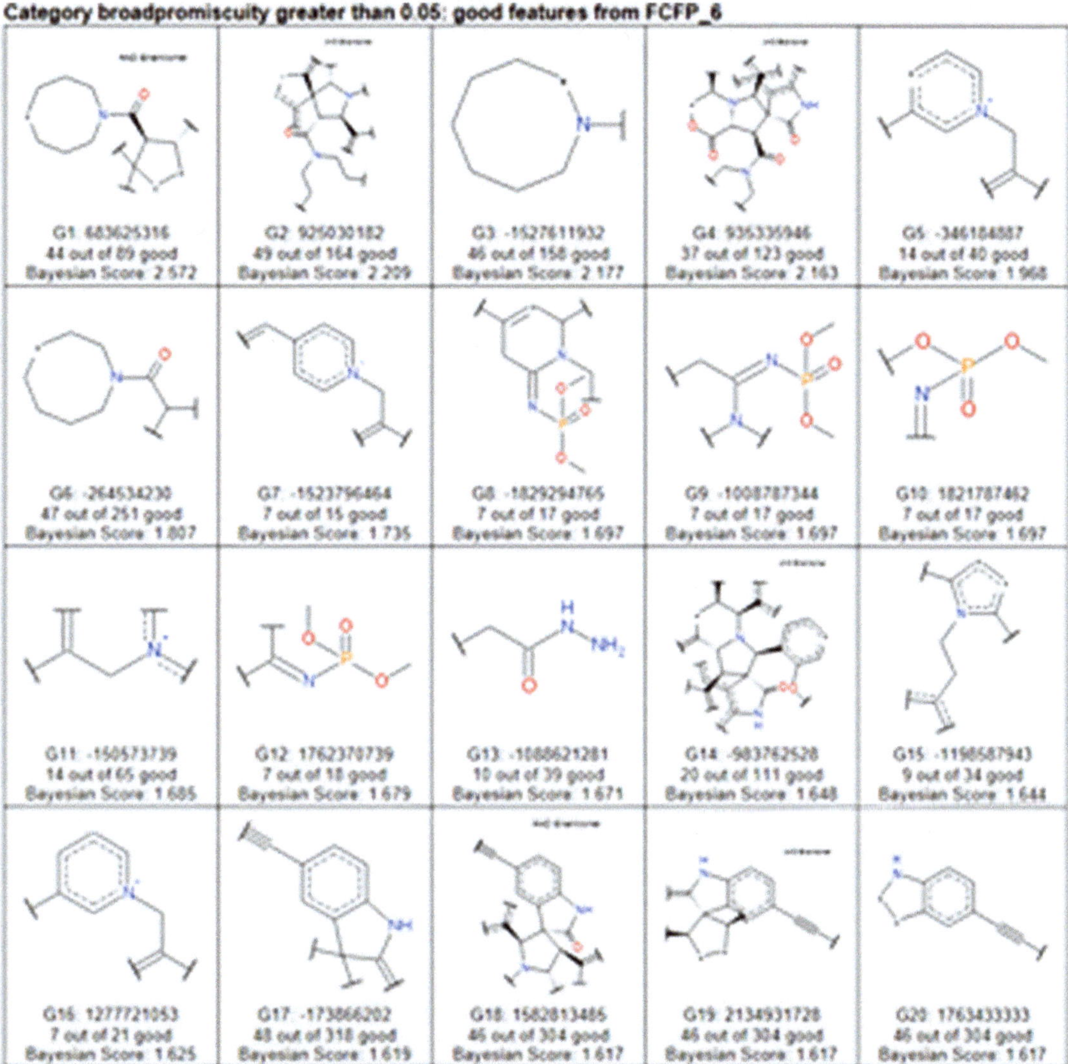

Fig. 10 (**a**) ~15,000 compounds with binding data to 100 different proteins good fingerprints. (**b**) ~15,000 compounds with binding data to 100 different proteins bad fingerprints. DOI: https:/doi.org/10.6084/m9.figshare.3206266

CDD Models within the CDD Vault, and will also enable academic and commercial models from multiple parties to be securely integrated without the need to share underlying data. We will likely add additional flexibility in terms of algorithms, descriptors and methods for assessing applicability of the models used. Providing CDD Models in the CDD Vault environment enables users to build models with private and or public data and make predictions which can be used to make decisions as to which molecules to make or buy. These technologies for data mining and modeling are within

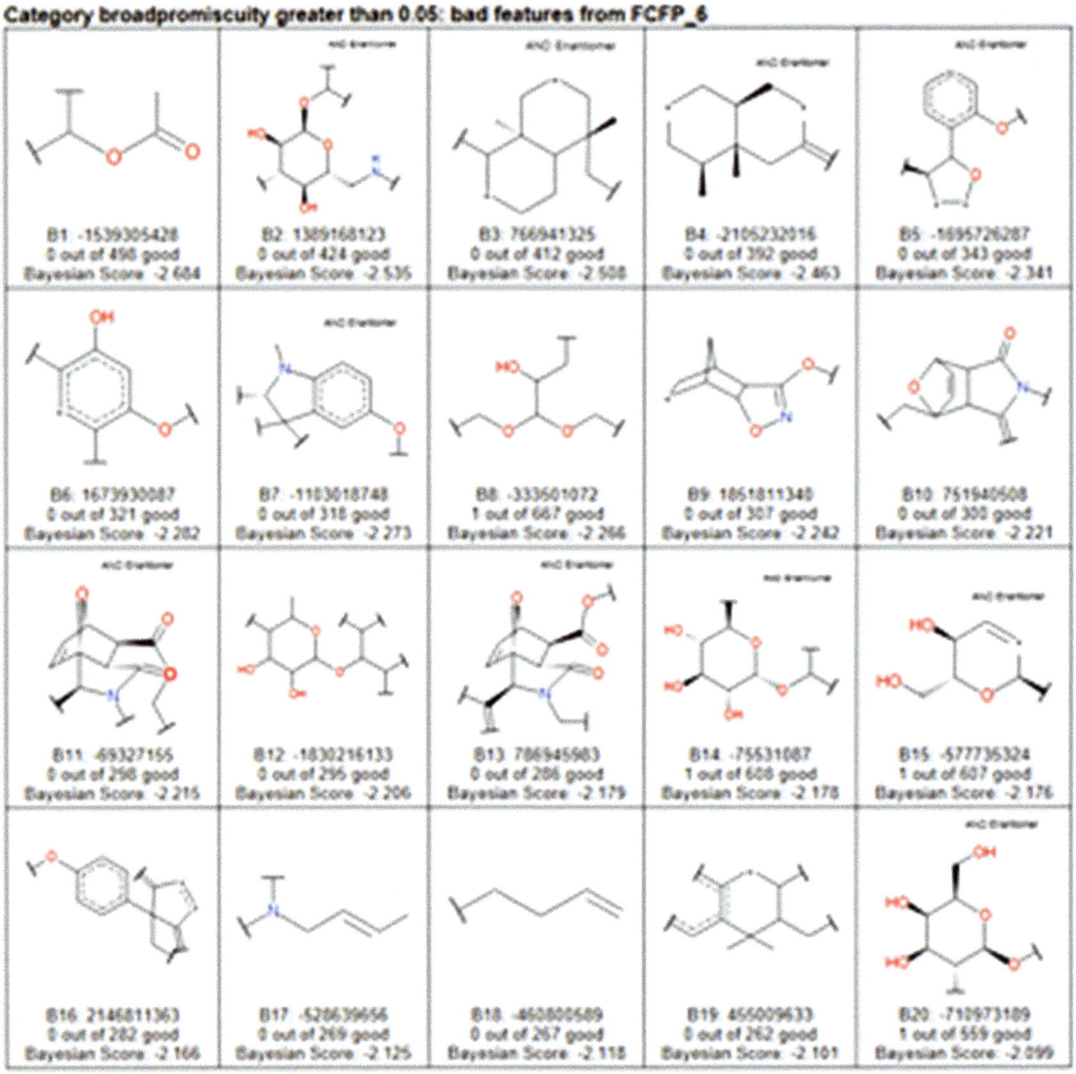

Fig. 10 (continued)

reach of non-cheminformatics experts in academia and small companies, which should help level the playing field for drug discovery.

Acknowledgments

We acknowledge that the Bayesian model software within CDD was developed with support from Award Number 9R44TR000942-02 "Biocomputation across distributed private datasets to enhance drug discovery" from the NIH NCATS. The CDD TB has been developed thanks to funding from the Bill and Melinda Gates

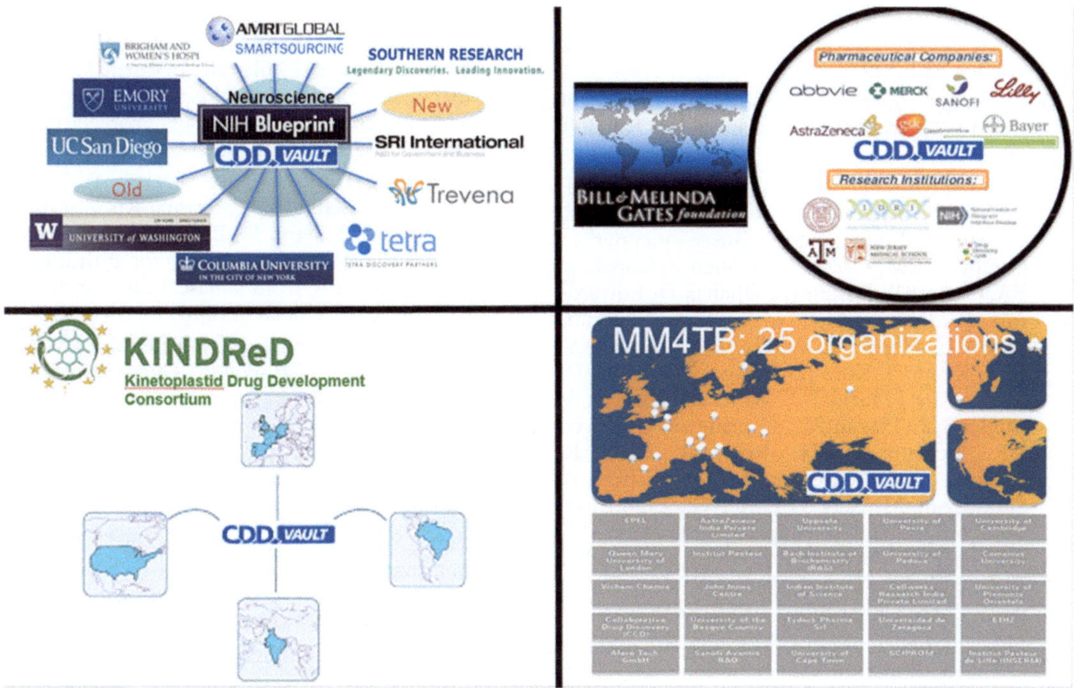

Fig. 11 Examples of Collaborative Drug Discovery Vault used in large public-private collaborations. DOI: https:/doi.org/10.6084/m9.figshare.3206266

Foundation (Grant#49852 "Collaborative drug discovery for TB through a novel database of SAR data optimized to promote data archiving and sharing"). The work was partially supported by a grant from the European Community's Seventh Framework Program (grant 260872, MM4TB Consortium) to S.E. S.E. gratefully acknowledges Biovia (formerly Accelrys) for providing Discovery Studio and Dr. Alexander Perryman and Dr. Joel Freundlich for their feedback and collaboration on CDD models. We sincerely acknowledge our many colleagues, collaborators, and advocates who have contributed to the development of CDD over the years.

References

1. Macarron R, Banks MN, Bojanic D, Burns DJ, Cirovic DA, Garyantes T, Green DV, Hertzberg RP, Janzen WP, Paslay JW, Schopfer U, Sittampalam GS (2011) Impact of high-throughput screening in biomedical research. Nat Rev Drug Discov 10:188–195
2. Ekins S, Waller CL, Bradley MP, Clark AM, Williams AJ (2013) Four disruptive strategies for removing drug discovery bottlenecks. Drug Discov Today 18:265–271
3. Oprea TI, Bologa CG, Boyer S, Curpan RF, Glen RC, Hopkins AL, Lipinski CA, Marshall GR, Martin YC, Ostopovici-Halip L, Rishton G, Ursu O, Vaz RJ, Waller C, Waldmann H, Sklar LA (2009) A crowdsourcing evaluation of the Nih chemical probes. Nat Chem Biol 5:441–447
4. Roy A, McDonald PR, Sittampalam S, Chaguturu R (2010) Open access high throughput drug discovery in the public domain: a Mount Everest in the making. Curr Pharm Biotechnol 11:764–778

5. Kaiser J (2011) National Institutes of Health. Drug-screening program looking for a home. Science 334:299
6. Frye S, Crosby M, Edwards T, Juliano R (2011) US academic drug discovery. Nat Rev Drug Discov 10:409–410
7. Arrowsmith CH, Audia JE, Austin C, Baell J, Bennett J, Blagg J, Bountra C, Brennan PE, Brown PJ, Bunnage ME, Buser-Doepner C, Campbell RM, Carter AJ, Cohen P, Copeland RA, Cravatt B, Dahlin JL, Dhanak D, Edwards AM, Frederiksen M, Frye SV, Gray N, Grimshaw CE, Hepworth D, Howe T, Huber KV, Jin J, Knapp S, Kotz JD, Kruger RG, Lowe D, Mader MM, Marsden B, Mueller-Fahrnow A, Muller S, O'Hagan RC, Overington JP, Owen DR, Rosenberg SH, Roth B, Ross R, Schapira M, Schreiber SL, Shoichet B, Sundstrom M, Superti-Furga G, Taunton J, Toledo-Sherman L, Walpole C, Walters MA, Willson TM, Workman P, Young RN, Zuercher WJ (2015) The promise and peril of chemical probes. Nat Chem Biol 11:536–541
8. Litterman N, Lipinski CA, Bunin BA, Ekins S (2014) Computational prediction and validation of an expert's evaluation of chemical probes. J Chem Inf Model 54:2996–3004
9. Payne DA, Gwynn MN, Holmes DJ, Pompliano DL (2007) Drugs for bad bugs: confronting the challenges of antibacterial discovery. Nat Rev Drug Discov 6:29–40
10. Wassermann AM, Camargo LM, Auld DS (2014) Composition and applications of focus libraries to phenotypic assays. Front Pharmacol 5:164
11. Mak PA, Rao SP, Ping Tan M, Lin X, Chyba J, Tay J, Ng SH, Tan BH, Cherian J, Duraiswamy J, Bifani P, Lim V, Lee BH, Ling Ma N, Beer D, Thayalan P, Kuhen K, Chatterjee A, Supek F, Glynne R, Zheng J, Boshoff HI, Barry CE 3rd, Dick T, Pethe K, Camacho LR (2012) A high-throughput screen to identify inhibitors of Atp homeostasis in non-replicating mycobacterium tuberculosis. ACS Chem Biol 7:1190–1197
12. Stanley SA, Grant SS, Kawate T, Iwase N, Shimizu M, Wivagg C, Silvis M, Kazyanskaya E, Aquadro J, Golas A, Fitzgerald M, Dai H, Zhang L, Hung DT (2012) Identification of novel inhibitors of M. tuberculosis growth using whole cell based high-throughput screening. ACS Chem Biol 7:1377–1384
13. Gold B, Pingle M, Brickner SJ, Shah N, Roberts J, Rundell M, Bracken WC, Warrier T, Somersan S, Venugopal A, Darby C, Jiang X, Warren JD, Fernandez J, Ouerfelli O, Nuermberger EL, Cunningham-Bussel A, Rath P, Chidawanyika T, Deng H, Realubit R, Glickman JF, Nathan CF (2012) Nonsteroidal anti-inflammatory drug sensitizes mycobacterium tuberculosis to endogenous and exogenous antimicrobials. Proc Natl Acad Sci U S A 109:16004–16011
14. Magnet S, Hartkoorn RC, Szekely R, Pato J, Triccas JA, Schneider P, Szantai-Kis C, Orfi L, Chambon M, Banfi D, Bueno M, Turcatti G, Keri G, Cole ST (2010) Leads for antitubercular compounds from kinase inhibitor library screens. Tuberculosis (Edinb) 90:354–360
15. Oprea TI, Matter H (2004) Integrating virtual screening in lead discovery. Curr Opin Chem Biol 8:349–358
16. Ekins S, Mestres J, Testa B (2007) In silico pharmacology for drug discovery: applications to targets and beyond. Br J Pharmacol 152:21–37
17. Ekins S, Mestres J, Testa B (2007) In silico pharmacology for drug discovery: methods for virtual ligand screening and profiling. Br J Pharmacol 152:9–20
18. McGaughey GB, Sheridan RP, Bayly CI, Culberson JC, Kreatsoulas C, Lindsley S, Maiorov V, Truchon JF, Cornell WD (2007) Comparison of topological, shape, and docking methods in virtual screening. J Chem Inf Model 47:1504–1519
19. Lombardo F, Obach RS, Dicapua FM, Bakken GA, Lu J, Potter DM, Gao F, Miller MD, Zhang Y (2006) A hybrid mixture discriminant analysis-random forest computational model for the prediction of volume of distribution of drugs in human. J Med Chem 49:2262–2267
20. Lombardo F, Obach RS, Shalaeva MY, Gao F (2004) Prediction of human volume of distribution values for neutral and basic drugs. 2. Extended data set and leave-class-out statistics. J Med Chem 47:1242–1250
21. Lombardo F, Obach RS, Shalaeva MY, Gao F (2002) Prediction of volume of distribution values in humans for neutral and basic drugs using physicochemical measurements and plasma protein binding. J Med Chem 45:2867–2876
22. Lombardo F, Shalaeva MY, Tupper KA, Gao F (2001) Elogdoct: a tool for lipophilicity determination in drug discovery. 2. Basic and neutral compounds. J Med Chem 44:2490–2497
23. Lombardo F, Blake JF, Curatolo WJ (1996) Computation of brain-blood partitioning of organic solutes via free energy calculations. J Med Chem 39:4750–4755
24. Lipinski CA, Lombardo F, Dominy BW, Feeney PJ (1997) Experimental and computational approaches to estimate solubility and

permeability in drug discovery and development settings. Adv Drug Deliv Rev 23:3–25

25. Ekins S, Ring BJ, Grace J, McRobie-Belle DJ, Wrighton SA (2000) Present and future in vitro approaches for drug metabolism. J Pharmacol Toxicol Methods 44:313–324

26. Huggett B (2016) Academic partnerships 2015. Nat Biotechnol 34:372

27. Zientek M, Stoner C, Ayscue R, Klug-McLeod J, Jiang Y, West M, Collins C, Ekins S (2010) Integrated in silico-in vitro strategy for addressing cytochrome P450 3a4 time-dependent inhibition. Chem Res Toxicol 23:664–676

28. Lombardo F, Shalaeva MY, Tupper KA, Gao F, Abraham MH (2010) Elogpoct a tool for lipophilicity determination in drug discovery. J Med Chem 43:2922–2928

29. Lagorce D, Sperandio O, Galons H, Miteva MA, Villoutreix BO (2008) Faf-Drugs2: free Adme/tox filtering tool to assist drug discovery and chemical biology projects. BMC Bioinformatics 9:396

30. Villoutreix BO, Renault N, Lagorce D, Sperandio O, Montes M, Miteva MA (2007) Free resources to assist structure-based virtual ligand screening experiments. Curr Protein Pept Sci 8:381–411

31. Ekins S (2007) Computational toxicology: risk assessment for pharmaceutical and environmental chemicals. John Wiley and Sons, Hoboken, NJ

32. Balani SK, Miwa GT, Gan LS, Wu JT, Lee FW (2005) Strategy of utilizing in vitro and in vivo Adme tools for lead optimization and drug candidate selection. Curr Top Med Chem 5:1033–1038

33. van De Waterbeemd H, Smith DA, Beaumont K, Walker DK (2001) Property-based design: optimization of drug absorption and pharmacokinetics. J Med Chem 44:1313–1333

34. Walters WP, Murcko MA (2002) Prediction of 'Drug-likeness'. Adv Drug Deliv Rev 54:255–271

35. ChEMBL. http://www.ebi.ac.uk/chembldb/index.php

36. Bento AP, Gaulton A, Hersey A, Bellis LJ, Chambers J, Davies M, Kruger FA, Light Y, Mak L, McGlinchey S, Nowotka M, Papadatos G, Santos R, Overington JP (2014) The Chembl bioactivity database: an update. Nucleic Acids Res 42:D1083–D1090

37. Gaulton A, Bellis LJ, Bento AP, Chambers J, Davies M, Hersey A, Light Y, McGlinchey S, Michalovich D, Al-Lazikani B, Overington JP (2012) Chembl: a large-scale bioactivity database for drug discovery. Nucleic Acids Res 40:D1100–D1107

38. Papadatos G, Overington JP (2014) The Chembl database: a taster for medicinal chemists. Future Med Chem 6:361–364

39. Wang Y, Xiao J, Suzek TO, Zhang J, Wang J, Bryant SH (2009) Pubchem: a public information system for analyzing bioactivities of small molecules. Nucleic Acids Res 37:W623–W633

40. Wang Y, Bolton E, Dracheva S, Karapetyan K, Shoemaker BA, Suzek TO, Wang J, Xiao J, Zhang J, Bryant SH (2010) An overview of the Pubchem Bioassay resource. Nucleic Acids Res 38:D255–D266

41. Huang R, Xia M, Sakamuru S, Zhao J, Shahane SA, Attene-Ramos M, Zhao T, Austin CP, Simeonov A (2016) Modelling the Tox21 10 K chemical profiles for in vivo toxicity prediction and mechanism characterization. Nat Commun 7:10425

42. Dix DJ, Houck KA, Martin MT, Richard AM, Setzer RW, Kavlock RJ (2007) The Toxcast program for prioritizing toxicity testing of environmental chemicals. Toxicol Sci 95:5–12

43. Shah F, Greene N (2014) Analysis of Pfizer compounds in Epa's Toxcast chemicals-assay space. Chem Res Toxicol 27:86–98

44. Hohman M, Gregory K, Chibale K, Smith PJ, Ekins S, Bunin B (2009) Novel web-based tools combining chemistry informatics, biology and social networks for drug discovery. Drug Discov Today 14:261–270

45. Ekins S, Hohman M, Bunin BA (2011) Pioneering use of the cloud for development of the collaborative drug discovery (Cdd) database. In: Ekins S, Hupcey MAZ, Williams AJ (eds) Collaborative computational technologies for biomedical research. Wiley and Sons, Hoboken, pp 335–361

46. Clark AM, Dole K, Coulon-Spector A, McNutt A, Grass G, Freundlich JS, Reynolds RC, Ekins S (2015) Open source Bayesian models: 1. Application to Adme/Tox and drug discovery datasets. J Chem Inf Model 55:1231–1245

47. Clark AM, Dole K, Ekins S (2015) Open source Bayesian models: 3. Composite models for prediction of binned responses. J Chem Inf Model 56:275–285

48. Clark AM, Ekins S (2015) Open source Bayesian models: 2. Mining a "big dataset" to create and validate models with Chembl. J Chem Inf Model 55:1246–1260

49. Balakin KV (2010) Pharmaceutical data mining : approaches and applications for drug discovery. John Wiley & Sons, Hoboken, NJ

50. Yan SF, King FJ, He Y, Caldwell JS, Zhou Y (2006) Learning from the data: mining of large high-throughput screening databases. J Chem Inf Model 46:2381–2395
51. Crisman TJ, Parker CN, Jenkins JL, Scheiber J, Thoma M, Kang ZB, Kim R, Bender A, Nettles JH, Davies JW, Glick M (2007) Understanding false positives in reporter gene assays: in silico chemogenomics approaches to prioritize cell-based Hts data. J Chem Inf Model 47:1319–1327
52. Johnson RL, Huang R, Jadhav A, Southall N, Wichterman J, MacArthur R, Xia M, Bi K, Printen J, Austin CP, Inglese J (2009) A quantitative high-throughput screen for modulators of Il-6 signaling: a model for interrogating biological networks using chemical libraries. Mol BioSyst 5:1039–1050
53. Hammann F, Drewe J (2012) Decision tree models for data mining in hit discovery. Expert Opin Drug Discov 7:341–352
54. Guilloux VL, Arrault A, Colliandre L, Bourg S, Vayer P, Morin-Allory L (2012) Mining collections of compounds with screening assistant 2. J Cheminform 4:20
55. Takada N, Ohmori N, Okada T (2013) Mining basic active structures from a large-scale database. J Cheminform 5:15
56. Soufan O, Ba-alawi W, Afeef M, Essack M, Rodionov V, Kalnis P, Bajic VB (2015) Mining chemical activity status from high-throughput screening assays. PLoS One 10:e0144426
57. Howe EA, de Souza A, Lahr DL, Chatwin S, Montgomery P, Alexander BR, Nguyen DT, Cruz Y, Stonich DA, Walzer G, Rose JT, Picard SC, Liu Z, Rose JN, Xiang X, Asiedu J, Durkin D, Levine J, Yang JJ, Schurer SC, Braisted JC, Southall N, Southern MR, Chung TD, Brudz S, Tanega C, Schreiber SL, Bittker JA, Guha R, Clemons PA (2015) Bioassay research database (Bard): chemical biology and probe-development enabled by structured metadata and result types. Nucleic Acids Res 43:D1163–D1170
58. Ekins S, Boulanger B, Swaan PW, Hupcey MA (2002) Towards a new age of virtual Adme/Tox and multidimensional drug discovery. Mol Divers 5:255–275
59. Gupta RR, Gifford EM, Liston T, Waller CL, Bunin B, Ekins S (2010) Using open source computational tools for predicting human metabolic stability and additional Adme/Tox properties. Drug Metab Dispos 38:2083–2090
60. Ekins S, Casey AC, Roberts D, Parish T, Bunin BA (2014) Bayesian models for screening and Tb mobile for target inference with mycobacterium tuberculosis. Tuberculosis (Edinb) 94:162–169
61. Ekins S, Reynolds RC, Franzblau SG, Wan B, Freundlich JS, Bunin BA (2013) Enhancing hit identification in mycobacterium tuberculosis drug discovery using validated dual-event Bayesian models. PLoS One 8:e63240
62. Ekins S, Reynolds RC, Kim H, Koo MS, Ekonomidis M, Talaue M, Paget SD, Woolhiser LK, Lenaerts AJ, Bunin BA, Connell N, Freundlich JS (2013) Bayesian models leveraging bioactivity and cytotoxicity information for drug discovery. Chem Biol 20:370–378
63. Ekins S, Freundlich JS, Hobrath JV, White EL, Reynolds RC (2014) Combining computational methods for hit to lead optimization in mycobacterium tuberculosis drug discovery. Pharm Res 31:414–435
64. Ekins S, Freundlich JS, Reynolds RC (2013) Fusing dual-event datasets for mycobacterium tuberculosis machine learning models and their evaluation. J Chem Inf Model 53:3054–3063
65. Ekins S, Pottorf R, Reynolds RC, Williams AJ, Clark AM, Freundlich JS (2014) Looking back to the future: predicting in vivo efficacy of small molecules versus mycobacterium tuberculosis. J Chem Inf Model 54:1070–1082
66. Ekins S, de Siqueira-Neto JL, McCall LI, Sarker M, Yadav M, Ponder EL, Kallel EA, Kellar D, Chen S, Arkin M, Bunin BA, McKerrow JH, Talcott C (2015) Machine learning models and pathway genome data base for Trypanosoma cruzi drug discovery. PLoS Negl Trop Dis 9:e0003878
67. Ekins S, Freundlich JS, Clark AM, Anantpadma M, Davey RA, Madrid P (2016) Machine learning models identify molecules active against the ebola virus in vitro. F1000Res 4:1091
68. Perryman AL, Stratton TP, Ekins S, Freundlich JS (2016) Predicting mouse liver microsomal stability with "pruned" machine learning models and public data. Pharm Res 33:433–449
69. Ekins S, Clark AM, Wright SH (2015) Making transporter models for drug-drug interaction prediction mobile. Drug Metab Dispos 43:1642–1645
70. Clark AM, Sarker M, Ekins S (2014) New target predictions and visualization tools incorporating open source molecular fingerprints for Tb mobile 2.0. J Cheminform 6:38
71. Lipinski CA, Litterman N, Southan C, Williams AJ, Clark AM, Ekins S (2015) The parallel worlds of public or commercial chemistry and biology data. J Med Chem 58:2068–2076
72. Jones DR, Ekins S, Li L, Hall SD (2017) Computational approaches that predict

metabolic intermediate complex formation with Cyp3a4 (+B5). Drug Metab Dispos 35:1466–1475

73. Metz JT, Johnson EF, Soni NB, Merta PJ, Kifle L, Hajduk PJ (2011) Navigating the kinome. Nat Chem Biol 7:200–202
74. Davis MI, Hunt JP, Herrgard S, Ciceri P, Wodicka LM, Pallares G, Hocker M, Treiber DK, Zarrinkar PP (2011) Comprehensive analysis of kinase inhibitor selectivity. Nat Biotechnol 29:1046–1051
75. Clemons PA, Bodycombe NE, Carrinski HA, Wilson JA, Shamji AF, Wagner BK, Koehler AN, Schreiber SL (2010) Small molecules of different origins have distinct distributions of structural complexity that correlate with protein-binding profiles. Proc Natl Acad Sci U S A 107:18787–18792
76. Ekins S, Litterman NK, Lipinski CA, Bunin BA (2016) Thermodynamic proxies to compensate for biases in drug discovery methods. Pharm Res 33:194–205
77. Anastassiadis T, Deacon SW, Devarajan K, Ma H, Peterson JR (2011) Comprehensive assay of kinase catalytic activity reveals features of kinase inhibitor selectivity. Nat Biotechnol 29:1039–1045
78. Norman RA, Toader D, Ferguson AD (2012) Structural approaches to obtain kinase selectivity. Trends Pharmacol Sci 33:273–278
79. Niijima S, Shiraishi A, Okuno Y (2012) Dissecting kinase profiling data to predict activity and understand cross-reactivity of kinase inhibitors. J Chem Inf Model 52:901–912
80. Uitdehaag JC, Verkaar F, Alwan H, de Man J, Buijsman RC, Zaman GJ (2012) A guide to picking the most selective kinase inhibitor tool compounds for pharmacological validation of drug targets. Br J Pharmacol 166:858–876
81. Burrill GS (2010) In: Fourth annual CDD community meeting, San Francisco
82. Paillard G, Cochrane P, Jones PS, van Hoorn WP, Caracoti A, van Vlijmen H, Pannifer AD (2016) The Elf Honest Data Broker: informatics enabling public-private collaboration in a precompetitive arena. Drug Discov Today 21:97–102
83. http://rarediseases.info.nih.gov/Resources/Rare_Diseases_Information.aspx http://rarediseases.info.nih.gov/Resources/Rare_Diseases_Information.aspx

Chapter 15

Intravital Imaging of Human Melanoma Cells in the Mouse Ear Skin by Two-Photon Excitation Microscopy

Nathan Y. Bentolila, Raymond L. Barnhill, Claire Lugassy, and Laurent A. Bentolila

Abstract

Noninvasive imaging of reporter gene expression by two-photon excitation (2PE) laser scanning microscopy is uniquely suited to perform dynamic and multidimensional imaging down to single-cell detection sensitivity in vivo in deep tissues. Here we used 2PE microscopy to visualize green fluorescent protein (GFP) as a reporter gene in human melanoma cells implanted into the dermis of the mouse ear skin. We first provide a step-by-step methodology to set up a 2PE imaging model of the mouse ear's skin and then apply it for the observation of the primary tumor and its associated vasculature in vivo. This approach is minimally invasive and allows repeated imaging over time and continuous visual monitoring of malignant growth within intact animals. Imaging fluorescence reporter gene expression in small living animals by 2PE provides a unique tool to investigate critical pathways and molecular events in cancer biology such as tumorigenesis and metastasis in vivo with high-spatial and temporal resolutions.

Key words Melanoma, Metastasis, Cancer cell migration, Angiotropism, Blood vessel, Dermis, Mouse ear, Two-photon microscopy, Multiphoton, Intravital imaging, Reporter gene expression

1 Introduction

Metastases, the development of secondary malignant growths at a distance from a primary site of cancer, are the main cause of cancer-associated death. How and why cancer cells leave their original location, travel throughout the body, and home on a new site to form new tumors has been difficult to observe in real time. That is in part because the various steps involved in the metastatic process are ephemeral events that are happening deep inside the body and also because tumor cells are usually difficult to distinguish from normal tissue.

Endowing tumor cells with fluorescence reporter gene expression enables direct observation of tumor growth and metastasis formation by various optical methodologies. Of particular importance, two-photon excitation (2PE) laser scanning microscopy (also

Robert Damoiseaux and Samuel Hasson (eds.), *Reporter Gene Assays: Methods and Protocols*, Methods in Molecular Biology, vol. 1755, https://doi.org/10.1007/978-1-4939-7724-6_15,

referred to as multiphoton microscopy) has expanded the application of fluorescence reporter gene expression deep into intact tissues [1] and has emerged as a powerful approach to image tumor cell behavior in vivo [2, 3]. 2PE is a nonlinear process in which two low-energy photons combine energy to induce a higher-energy electronic transition in a fluorescent molecule [4]. The near-simultaneous absorption of two photons of light of roughly half the energy is usually achieved with an ultrafast, mode-locked Ti: Sapphire pulsed laser emitting in the near IR region of the spectrum (690–1040 nm). This excitation process has a number of unique advantages. First, the localization of fluorescence excitation is confined within a femtoliter size 3D focal volume [5]. The localized excitation volume also provides innate confocality without the need of a spatial confocal pinhole in front of the detector. This limits the out-of-focus background excitation which greatly improves image contrast and reduces photobleaching. Secondly, the use of deep red and near IR excitation wavelength afford deeper penetration into scattering samples allowing imaging of thicker biological specimen compared with conventional one-photon confocal microscopy. For all these reasons, 2PE microscopy is especially valuable for noninvasive deep imaging of live tissues.

The mouse ear is relatively thin and flat. Its skin offers an ideal location to perform 2PE intravital microscopy to follow stable fluorescent protein reporter expression in dynamic processes from immune responses to tumor development [6, 7]. Here, we describe the use of the mouse ear skin in an experimental model of orthotopic injection of human melanoma cells with stable green fluorescent protein (GFP) expression. We present a general method to immobilize the mouse ear for imaging both the tumor and its associated vasculature in vivo in an undisturbed natural context with single cell resolution. This noninvasive and highly selective imaging of growing tumors, made possible by GFP fluorescence reporter expression, enables the tracking of tumor growth and metastasis formation in transplanted animals. 2PE microscopy imaging should accelerate the validation of cell-based reporter gene approaches for translation into preclinical small animal models.

2 Materials

All reagents are of the best analytical grade and all solutions should be prepared in Milli-Q water (resistivity of more than18 MΩ-cm).

2.1 Reagents, General Supplies and Equipment

1. DyLight® 594 *Lycopersicon esculentum* (Tomato) Lectin, (Vector Laboratories, Burlingame, CA, cat. no. DL-1177).
2. 0.5 mL Lo-Dose™ Insulin Syringe U-100 28G$^{1}/_{2}$ Micro-Fine™ IV Needle (0.36 × 13 mm), (BD Becton Dickinson, Franklin Lakes, NJ, cat. no. 329465).

3. WillCo-dish® Glass Bottom dishes (WillCo Wells B.V., Amsterdam, The Netherlands, cat. no. GWSB-5040).
4. MakerBot Replicator Desktop 3D Printer (MakerBot® Industries, LLC Brooklyn, NY USA).
5. UV handheld lamp.
6. Curved splinter forceps.
7. Cotton-tipped applicators (Covidien, cat. no. 8884540400).
8. Kimwipes delicate task wipes (Kimtech Science, cat. no. 34155).
9. Water bath and heating mat set for 37 °C.

2.2 Cells and Cell Culture Reagents

1. C8161 GFP human melanoma cell line [8].
2. Dulbecco's modified Eagle's medium (DMEM-500 mL, GIBCO® by Life Technologies, cat. no. 11995-065) supplemented with 10% (50 mL) inactivated fetal calf serum (Gemini Bio-products, Sacramento, CA, cat. no. 100-106), 5 mL of L-glutamine 200 mM (100×, GIBCO® by Life Technologies, cat. no. 25030-081), 5 mL of penicillin–streptomycin (10,000 units, GIBCO® by Life Technologies, cat. no. 15140-122) and 4 μL of 2-β-mercaptoethanol at 14.3 M (Sigma-Aldrich, cat. no. M7522).
3. Dulbecco's Phosphate-Buffered Saline (DPBS) without Ca^{2+} and Mg^{2+} (1×, DPBS-500 mL, GIBCO® by Life Technologies, cat. no. 14190-144).
4. 1× Solution of 0.25% trypsin–EDTA (GIBCO® by Life Technologies, cat. no. 25200-056).
5. Vented 75 cm^2–250 mL Falcon tissue culture flasks (Fisher Scientific, cat. no. 13-680-65).
6. 15 mL Falcon tubes with conical bottoms (Fisher Scientific, cat. no. 14-959-49B).
7. 50 mL Falcon tubes with conical bottoms (Fisher Scientific, cat. no. 14-432-22).
8. Plastic Pasteur pipettes (Celltreat Scientific Products, Shirley, MA, cat. no. 229285).
9. Cell counter and phase hemacytometer (VWR international, cat no. 15170-263).
10. CO_2 cell culture incubator (SANYO, cat. no. MCO-36AIC) set up at 37 °C.
11. Tissue culture hood biosafety cabinet (The Baker Company, SterilGARD III, cat. no. SG-403a).
12. Fluorescence microscope (Leica Microsystems Inc., Exton, PA, model no. DMIL) equipped with a GFP filter cube (Excitation BP 450–490 nm, Dichroic RSP 510 nm, Emission LP 515 nm).

2.3 Experimental Animals and Reagents

1. Athymic nude, nu/nu mouse strain JAX 002019 NU/J (The Jackson Laboratory, Bar Harbor, ME) (*see* **Note 1**).
2. Artificial tears (Lubricant Ophthalmic Ointment), (Akorn Animal Health, Lake Forest, IL, cat. no. 17478-162-35).
3. Nair (Hair remover lotion), available in any drugstore and pharmacy.
4. Isoflurane, USP (Isothesia-250 mL, cat. no. 029405, Henry Schein® Animal Health (*see* **Note 2**).
5. Oxygen supply (provided by a central distribution system or by pressurized tanks).
6. Anesthetic vaporizer (Summit Medical Equipment Company, Bend, OR).

2.4 Two-Photon Excitation (2PE) Laser Scanning Microscopy

1. Confocal upright microscope with fixed stage (Leica Microsystems Inc., Exton, PA, model no. DM6000 CFS).
2. Tunable (690–1040 nm) Ti:Sapphire pulsed laser providing >2.4 W average power and <70 fs pulse width (Spectra-Physics, Santa Clara, CA, model no. Mai Tai® DeepSee™).
3. High-sensitivity nondescanned (NDD) detectors (Leica Microsystems Inc., Exton, PA, model no. HyD SMD).
4. Water immersion objective (Leica HC PL IRAPO 20×/0.75).
5. Dual-band filter set for NDD detection of GFP and DyLight® 594: BP 525/50 nm (GFP), Dichroic RSP 560 nm, BP 624/40 nm (DyLight® 594) (Leica Microsystems Inc., Exton, PA).
6. 4′ × 6′ optical table (TMC, Peabody, MA).

2.5 Intravital Ear Stage Platform

1. Custom-built 3D-printed ear-imaging stage platform [9] (*see* **Note 3**).
2. Heating pad and temperature controller set for 37 °C (Watlow, Riverside, CA).

3 Methods

Please note that all experiments using laboratory animals should be conducted in accordance with the relevant ethics guidelines and regulations in place at your institution.

3.1 Cell Culture

1. Subculture cells in vented 75 cm^2 tissue culture flasks in order to achieve 70–80% confluence of actively growing cells (Fig. 1a).
2. Remove medium with a Pasteur pipette and wash the cells with 10 mL of DPBS.

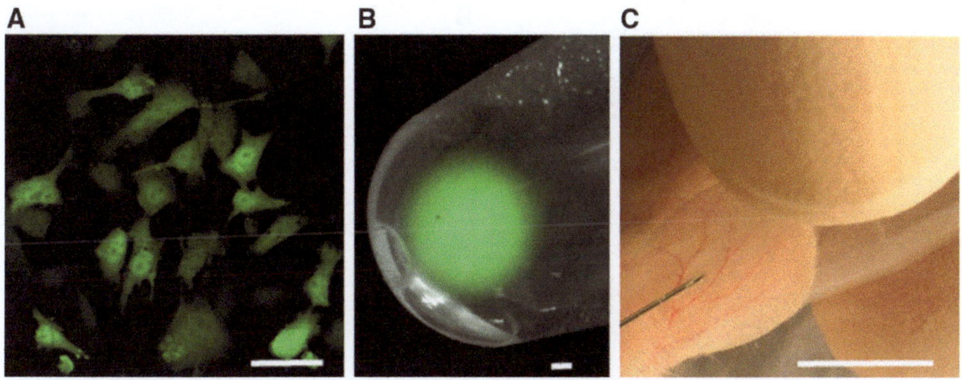

Fig. 1 Procedure for intradermal injection of human melanoma cells in the mouse ear. (**a**) Human C8161 melanoma cells expressing GFP are grown into a semiconfluent monolayer. Scale bar, 50 μm. (**b**) After trypsinization, C8161 cells are centrifuged and collected in a 1.5 mL microcentrifuge tube. Scale bar, 100 μm. (**c**) While the mouse is anesthetized and laying onto a heating pad, the conical end of 1.5 mL microcentrifuge tube is inserted inside the ear. 10–20 μL of the cell suspension is loaded into an insulin syringe equipped with a 28G$^1/_2$ needle. The needle is held as flat as possible bevel up and the cells are injected into the dermis of the anesthetized mouse. A transient bulge should be visible immediately after the injection. Scale bar, 1 cm for **(c)**

3. Add 0.5 mL of trypsin to detach cells and incubate 4 min at 37 °C. Monitor the enzymatic digestion with a light microscope.
4. Stop digestion by adding 10 mL of fresh complete medium. Transfer the cells into a 15 mL tube.
5. Count cells with hemocytometer.
6. Centrifuge the solution at room temperature for 10 min at 110 × *g*.
7. Discard the supernatant by aspirating it off carefully with a Pasteur pipette (Fig. 1b).
8. Resuspend the cell pellet in 20–100 μL of DPBS. Adjust final volume to achieve a cell concentration of about 5000 cells per μL (*see* **Note 4**).
9. Fill an insulin syringe with up to 10–20 μL of the cell suspension to be injected in the ear.

3.2 Intradermal Injection in the Mouse Ear

1. Anesthetize the mouse using a mixture of 4–5% isoflurane and oxygen under a steady flow (*see* **Note 2**).
2. Place the mouse on a heating pad set to 37 °C to maintain its body temperature at 37 °C throughout the procedure (*see* **Note 5**).
3. With a cotton-tip applicator, gently apply artificial tears to both eyes to prevent dryness.
4. While the mouse is lying flat on its belly, insert the conical end of a 1.5 mL microcentrifuge tube inside the ear such that the mouse ear is wrapped flat against the tube wall (Fig. 1c).

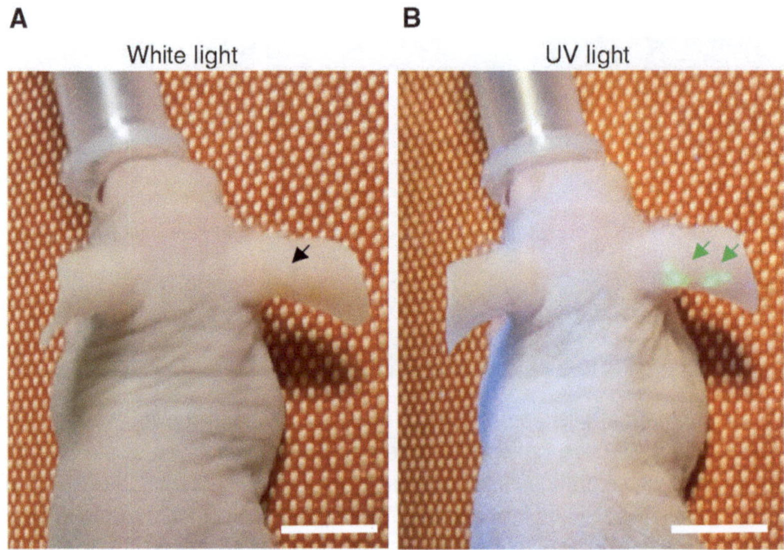

Fig. 2 Tumors arising from the injection of C8161 human melanoma cells after 2 weeks. (**a**) Macroscopic bright-field image of site of injection (black arrow). (**b**) Fluorescence image showing the formation of tumors under the ear skin (green arrows). Scale bars, 1 cm

5. Once the mouse ear is properly and firmly positioned, gently rest the needle of the insulin syringe on the mouse ear with the bevel up (Fig. 1c).
6. Carefully pierce the mouse skin while maintaining the needle bevel up as flat as possible, almost parallel to the ear (*see* **Note 6**).
7. Slowly inject all or part of the syringe content.
8. A transient bulge should be visible immediately after the injection.
9. Slowly remove the needle to minimize the loss of some of the injected cell suspension.
10. Allow the mouse to recover on the heating pad.
11. Imaging can start on day 1 after injection and/or whenever desired thereafter.
12. Tumor growth under the ear skin can be monitored with a UV hand-held lamp (Fig. 2).

3.3 Preimaging Preparation of the Mouse Ear

1. If labeling of the blood vessels is desired (optional), red lectin can be injected in the vein at the base of the dorsal side of the mouse ear (Fig. 3a) using an insulin syringe and the microcentrifuge tube holding method described above (Fig. 1c) (*see* **Note 7**).
2. Removal of the hair on the mouse ear can also be performed at this point to facilitate later imaging by 2PE (optional for nude mice).

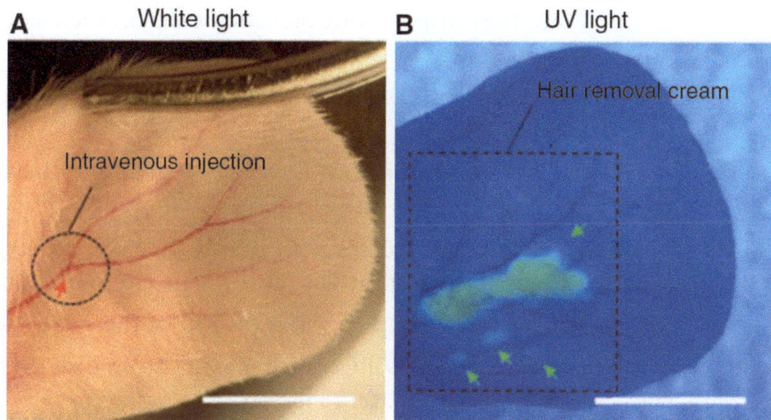

Fig. 3 Preimaging preparation of the mouse ear. (**a**) Macroscopic bright-field image of the vasculature of the mouse ear showing the site of Dylight594-Tomato Lectin intravenous injection to label blood vessels (red arrow). (**b**) Close-up view of the mouse ear showing the green melanoma tumors under the skin upon UV exposure (green arrows). The dotted rectangle denotes the area to apply cream for hair removal before 2PE imaging. Scale bars, 5 mm

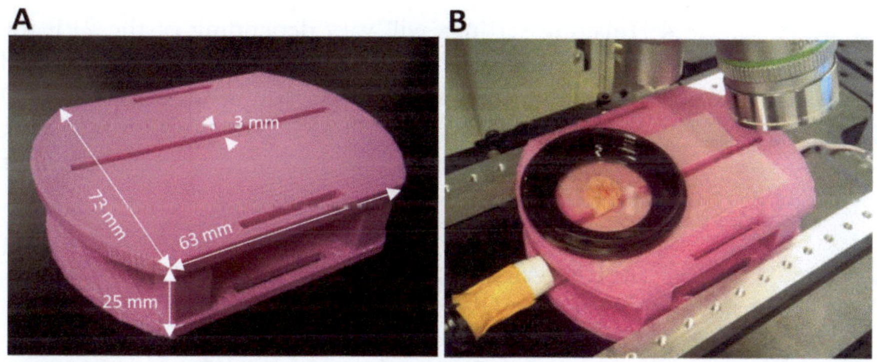

Fig. 4 Procedure for immobilizing the mouse ear for intravital imaging. (**a**) 3D-printed plastic mouse ear stage. (**b**) Photograph of the 2PE imaging setup with the mouse ear fastened onto the stage. The ear is gently pulled through one of the slits and secured in place with double-sided tape. A drop of water is placed on top of the ear to prevent dehydration before affixing a 50 mm WillCo-dish glass bottom dish that provides a flat surface for imaging and a reservoir for the 20× water objective. The objective is then lowered in the water-filled viewing chamber. The heating pad sits directly under the plastic stage insert

3. Using a PBS premoistened cotton-tip applicator, carefully apply the hair removal cream to the dorsal side of the mouse ear (Fig. 3b).
4. After 1–2 min, gently remove the cream with a wet cotton-tip applicator (*see* **Note 8**).

3.4 Mouse Ear Mounting onto the Imaging Platform

1. The custom 3D-printed stage holder (Fig. 4a) is placed directly on top of the 37 °C heating pad on the microscope stage (*see* **Note 3**).

2. Place two strips of double-sided tape on each side of the slit of the ear stage (Fig. 4b).
3. Carefully position the anesthetized mouse in the 3D-printed holder.
4. Maintain a continuous flow of low dose anesthesia (2% isoflurane and oxygen) (*see* **Note 5**).
5. Using flat curved forceps, gently pull the mouse ear through the slit and secure it flat on the tape (*see* **Note 9**).
6. Place a small drop of water on top of the ear to keep it moist.
7. Place a 50 mm WillCo-dish glass bottom dish on top of the ear and fill it with water (Fig. 4b).

3.5 Intravital 2PE Imaging

1. Carefully lower the 20× water objective into the water-filled viewing chamber (*see* **Note 10**).
2. Keep monitoring anesthesia and mouse body temperature closely (*see* **Note 11**).
3. Identify the location of the tumor and start imaging by turning on live preview.
4. Imaging settings will vary depending of the 2PE microscope used. We typically use the following settings on a 2PE Leica DM6000 CFS microscope equipped with two NDD HyD detectors: 832 pixels × 832 pixels resolution; 24 kHz scan frequency (six lines and 20 frames averaging); digital zoom 1–4×; 169 μm × 169 μm scan field area.
5. Sequentially scan the two fluorophores with the appropriate 2PE wavelength: 910 nm for GFP and 800 nm for DyLight® 594 Lectin (Fig. 5).

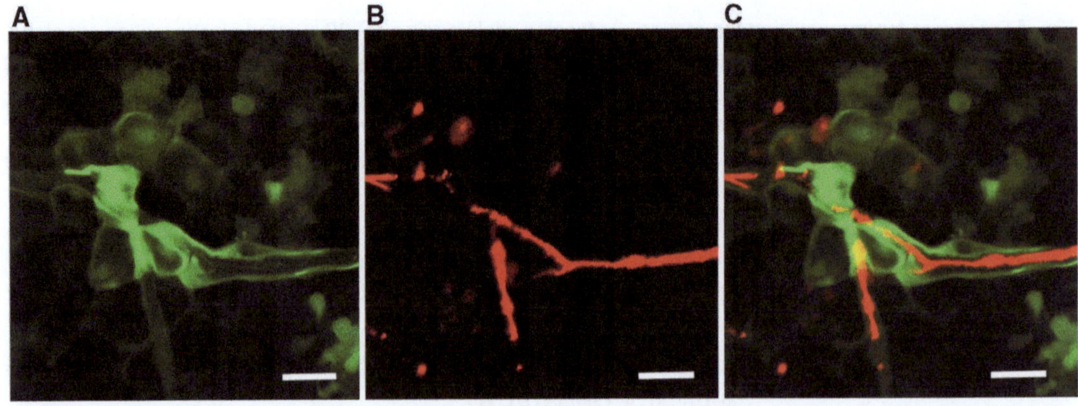

Fig. 5 Two-color intravital imaging of melanoma tumors in the mouse ear by 2PE microscopy. (**a**) Optical section through about 70 μm of dermis showing melanoma tumor cells expressing GFP. (**b**) Blood vessels labeled with Dylight594 Tomato Lectin. (**c**) Overlay image of GFP and Tomato Lectin. In these figures, it is possible to observe a clear angiotropism of individual melanoma cells spreading along the vessels [10]. Scale bars, 50 μm

6. Imaging can be performed for several hours or days later.
7. At the end of the imaging session, gently remove the WillCo-dish glass bottom dish and the mouse ear with a moist cotton-tip applicator.
8. If the intent is the keep the mouse for further imaging, let it recover on the heating-pad.
9. If it is to be euthanized, follow official regulations and guidelines approved by your institution.

4 Notes

1. Maintenance and surgical procedures were consistent with UCLA Animal Research Council Guidelines.
2. Federal law restricts this drug to use by or on the order of a licensed veterinarian.
3. The custom-built 3D-printed ear-imaging stage platform serves to shield the imaging from artifacts induced by the breathing motion and heartbeat of the mouse [9].
4. It is especially important to obtain a single-cell suspension as any remaining cell aggregates will clog the injection needle.
5. We recommend monitoring the mouse body temperature during and after anesthesia using an animal rectal probe (World Precision Instruments, cat. no. RET-3) as it can drop very quickly.
6. If the injection angle is too steep, you may risk piercing through both sides of the ear.
7. Intravenous injections can also be administered via the tail vein but higher doses of lectin are needed.
8. Always apply a minimal amount of the depilatory cream for the shortest period of time that is necessary to satisfactorily remove the excess hair. The ear skin is extremely thin and leaving the hair cream for too long could cause inflammation and damage to the skin. The cream must be removed thoroughly but gently with a wet cotton-tip applicator.
9. It is essential to move the body of the mouse at the same time the ear is pulled through the slit to avoid tearing apart the thin skin. Also, great care should be taken to make sure the ear lays flat on the tape (no folds) to make a flat surface for the coverslip.
10. Periodically check the water level as it will eventually evaporate over time.

11. When imaging for several hours, it is imperative to monitor the levels of both oxygen and isoflurane in the vaporizer as not to run out of anesthesia and have to deal with an awakening mouse in the middle of an experiment.

Acknowledgment

We thanks Dr. Danny Welch (The Kansas University Medical Center) for providing the C8161 GFP human melanoma cell line. 2PE microscopy was performed at the California NanoSystems Institute (CNSI) Advanced Light Microscopy/Spectroscopy Shared Resource Facility at UCLA with support from a NIH/National Center for Advancing Translational Science UCLA CTSI Grant (UL1TR000124). The authors also thank Mr. Ian Arenas from the YULA Genesis Innovation Lab for help with the 3D printing of the mouse ear stage.

References

1. Denk W, Delaney KR, Gelperin A, Kleinfeld D, Strowbridge BW, Tank DW, Yuste R (1994) Anatomical and functional imaging of neurons using 2-photon laser scanning microscopy. J Neurosci Methods 54:151–162
2. Beerling E, Ritsma L, Vrisekoop N, Derksen PW, van Rheenen J (2011) Intravital microscopy: new insights into metastasis of tumors. J Cell Sci 124:299–310
3. Condeelis J, Weissleder R (2010) In vivo imaging in cancer. Cold Spring Harb Perspect Biol 2 (12):a003848
4. Denk W, Strickler JH, Webb WW (1990) Two-photon laser scanning microscopy. Science 248(4951):73–76
5. Zipfel WR, Williams RM, Webb WW (2003) Nonlinear magic: multiphoton microscopy in the biosciences. Nat Biotechnol 21:1369–1377
6. Li JL, Goh CC, Keeble JL, Qin JS, Roediger B, Jain R, Wang Y, Chew WK, Weninger W, Ng LG (2012) Intravital multiphoton imaging of immune responses in the mouse ear skin. Nat Protoc 7(2):221–234
7. Chan KT, Jones SW, Brighton HE, Bo T, Cochran SD, Sharpless NE, Bear JE (2013) Intravital imaging of a spheroid-based orthotopic model of melanoma in the mouse ear skin. Intravital 2(2):e25805
8. Welch DR, Bisi JE, Miller BE, Conaway D, Seftor EA, Yohem KH, Gilmore LB, Seftor REB, Nakajima M, Hendrix MJC (1991) Characterization of a highly invasive and spontaneously metastatic human malignant melanoma cell line. Int J Cancer 47(2):227–237
9. Bentolila NY, Arenas I, Bentolila LA (2018) 3D-printed universal ear stage holder for two-photon microscopy imaging in living animals. Fully 3D-printable files are available at alms.cnsi.ucla.edu
10. Lugassy C, Zadran S, Bentolila LA, Wadehra M, Prakash R, Carmichael TS, Kleinman H, Péault B, Larue L, Barnhill RL (2014) Angiotropism, pericytic mimicry and extravascular migratory metastasis in melanoma: an alternative to intravascular cancer dissemination. Cancer Microenviron 7 (3):139–152

Index

Robert Damoiseaux and Samuel Hasson (eds.), *Reporter Gene Assays: Methods and Protocols*, Methods in Molecular Biology, vol. 1755, https://doi.org/10.1007/978-1-4939-7724-6,

MIX
Papier aus verantwortungsvollen Quellen
Paper from responsible sources
FSC® C105338

If you have any concerns about our products,
you can contact us on
ProductSafety@springernature.com

In case Publisher is established outside the EU,
the EU authorized representative is:
Springer Nature Customer Service Center GmbH
Europaplatz 3, 69115 Heidelberg, Germany

Printed by Libri Plureos GmbH
in Hamburg, Germany